SPECIAL RELATIVITY:
Applications to Particle Physics and the Classical Theory of Fields

ELLIS HORWOOD SERIES IN PHYSICS AND ITS APPLICATIONS

Series Editors: JOHN W. MASON, Scientific and Technical Consultant; MALCOLM COOPER, Professor, Department of Physics, University of Warwick; and E. H. GRANT, Department of Physics, King's College, University of London

Balas, J. and Szabo, V.	**HOLOGRAPHIC INTERFEROMETRY IN EXPERIMENTAL MECHANICS**
Choudhury, M.H.	**ELECTROMAGNETISM**
Delaney, C.F.G.	**ELECTRONICS FOR THE PHYSICIST WITH APPLICATIONS**
Dobrzynski, L., Blinowski, K. and Cooper, M.	**NEUTRONS AND SOLID STATE PHYSICS**
Dyson, N.A.	**AN INTRODUCTION TO NUCLEAR PHYSICS, WITH APPLICATIONS IN MEDICINE AND BIOLOGY**
Dyson, N.A.	**RADIATION PHYSICS WITH APPLICATIONS IN MEDICINE AND BIOLOGY**
Elwell, D.	**PHYSICS FOR ENGINEERS AND SCIENTISTS**
Gough, W., Richards, J.P.G. and Williams, R.P.	**VIBRATIONS AND WAVES**
Granier, R. and Gambini, D.-J.	**APPLIED RADIATION BIOLOGY AND PROTECTION**
Hasnain, S.S. (editor)	**SYNCHROTRON RADIATION AND BIOPHYSICS**
Hill, C.R. (editor)	**PHYSICAL PRINCIPLES OF MEDICAL ULTRASONICS**
Ignatowicz, I. and Kobendza, A.	**SEMICONDUCTING THIN FILMS OF A^2B^4 COMPOUNDS**
Martin, J.L.	**GENERAL RELATIVITY: A Guide to its Consequences for Gravity and Cosmology**
Rosser, W.G.V.	**AN INTRODUCTION TO STATISTICAL PHYSICS**
Rowlands, G.	**NON-LINEAR PHENOMENA IN SCIENCE AND ENGINEERING**
Saleem, M. and Rafique, M.	**SPECIAL RELATIVITY: Applications to Particle Physics and the Classical Theory of Fields**
Scott, V.D. and Love, G. (editors)	**QUANTITATIVE ELECTRON-PROBE MICROANALYSIS**
Steward, E.G.	**FOURIER OPTICS: An Introduction, 2nd Edition**
Wadas, R.	**BIOMAGNETISM**
Whorlow, R.W.	**RHEOLOGICAL TECHNIQUES, 2nd Edition**

SPECIAL RELATIVITY:
Applications to Particle Physics and the Classical Theory of Fields

MOHAMMAD SALEEM
Centre for High Energy Physics, University of the Punjab,
Lahore, Pakistan
MUHAMMAD RAFIQUE
Department of Mathematics, University of the Punjab, Lahore,
Pakistan

ELLIS HORWOOD
NEW YORK LONDON TORONTO SYDNEY TOKYO SINGAPORE

First published in 1992 by
ELLIS HORWOOD LIMITED
Market Cross House, Cooper Street,
Chichester, West Sussex, PO19 1EB, England

A division of
Simon & Schuster International Group
A Paramount Communications Company

Printed and bound in Great Britain
by Redwood Press, Melksham

British Library Cataloguing in Publication Data

A catalogue record for this book is available from the British Library

ISBN 0–13–827106–2

Library of Congress Cataloging-in-Publication Data

Available from the publisher

CONTENTS

PREFACE

Although a number of books have been written on the special theory of relativity, we still strongly feel that some very important aspects and applications of this branch of physics have been constantly ignored. The existing books generally give a brief survey of relativistic mechanics and electromagnetism after considering the conventional applications of the Lorentz transformation. We have noticed that from reading these books on relativity one cannot duly realize the importance of relativistic aspects in binary collisions. The wide spectrum of this topic appears to have attracted little attention. Moreover, group theoretical concepts which play a very significant role in present mainstream physics do not get their due share of attention in the treatment of the subject. Also, the role of 4-dimensional tensor technique in the development of electromagnetism has not been given proper emphasis.

In this book, as well as explaining the basic postulates of special relativity, we have discussed in detail and critically considered their influence in the fields of mechanics and electromagnetism. In relativistic mechanics, which has been developed by using 4-dimensional tensor technique, the approach is such that students intending to do research in high energy physics neither face any difficulty nor probably require any other book on relativity for understanding this topic. The relativistic expressions for the Lagrangian and the Hamiltonian of a charged particle have been determined, and the collision and decay of elementary particles have been discussed in detail.

Although the special theory of relativity has brought a revolutionary change in the field of mechanics where some new and highly unexpected results have emerged, the tensor formulation of electromagnetism whose basic equations remain Lorentz invariant has played a very important role in synthesizing the various physical concepts. We have also fully utilized the latter idea. Therefore, we do not have any hesitation in stating that anyone using field theory in any branch of physics will find this book highly beneficial.

The relativistic kinematics of binary collisions which, in most of the books on special relativity, is either omitted or is mentioned only as a passing remark has been given exhaustive treatment in an independent chapter.

The importance of the Poincaré group is now well recognized. Starting *ab initio*, we have thrown light on various kinds of Lorentz transformations and their subsets which emerge as special cases of this group, and have tried to illustrate, wherever possible, their physical significance. It is important to note, and it has been duly emphasized in this book, that one of the basic postulates of special relativity, stating that the laws of physics have the same form in all inertial frames of reference, is valid only under proper Lorentz transformations.

Superluminal particles – **tachyons** – have been discussed in the last chapter of the book in such a way as to arouse curiosity to investigate various aspects of this novel idea.

To make the book self-sufficient, appendices have been added so that those who have not studied tensor analysis or group theory should not have any difficulty in understanding the text.

We hope this book will be highly useful for students of physics and applied mathematics.

Mohammad Saleem
Muhammad Rafique

1

GALILEAN RELATIVITY AND ABSOLUTE MOTION IN SPACE

The special theory of relativity was propounded by Einstein in 1905. In order to understand the circumstances which led to the advent of this theory, we shall begin with an examination of Newton's first and second laws of motion in their proper perspective, and show how the failure of the attempts to determine absolute motion in space led to a dilemma in classical physics.

Newton's first law of motion states that a material object, if left to itself, continues in its state of rest or of uniform motion in a straight line. For this law to be meaningful we have to specify the reference body relative to which the motion of the material object is described. For instance, a man sitting in a train travelling with a speed of 70 km per hour along a straight track, will be at rest relative to the train but will be moving with a velocity of 70 km per hour with respect to the ground. Thus the description of the state of an object depends on the surroundings with respect to which it is specified. These surroundings themselves can be specified in terms of the rigid bodies around. Any one of these rigid bodies, supposed to be extended in all space, together with a time-recording device, a clock, is said to constitute a **reference system** or a **frame of reference.** All measurements in space and time may be made with respect to such frames. In actual practice, for the purpose of space measurement, we attach a coordinate system to the rigid body. In what follows, the location of an object shall always be specified by using **rectangular Cartesian coordinate systems.**

We have shown that the first law of motion has a meaning only with respect to a frame of reference. But is this law valid in all frames of reference? To answer this question, consider a man sitting in a bus. Whenever the bus is accelerated, the man receives a jerk and has a tendency to move although no force, extraneous to the frame, is acting on him. Thus Newton's first law does not hold good in accelerated frames of reference. A frame of reference in which Newton's first law of motion, also called the law of inertia, holds, is said to be an **inertial frame**. We shall *assume* that such frames do exist in nature. The question then arises, how would it be possible to find out whether a given frame of reference is inertial or not? For an inertial frame, an object, at rest or moving with uniform velocity in it, should not change its state in the absence of an external force. But, in practice, the gravitational forces are always present to influence the behaviour of the object under consideration and it is not possible to check what would happen if these forces were not present. However, a frame of reference in which the first law holds at least approximately can be determined by considering the behaviour of an object far removed from all other bodies. For instance, if we use a frame of reference which is rigidly attached to the fixed stars, then the position of the heavenly bodies relative to this frame is very nearly uniform, indicating that this frame is approximately an inertial frame of reference. The departure from uniformity can reasonably be accounted for as due to the influence of the stars upon one another.

We shall now show that a frame of reference which is in uniform motion relative to an inertial frame is also inertial. For this purpose let us first find the relation connecting two frames of reference moving uniformly with respect to each other.

The Galilean Transformation

Consider a frame of reference S′ which is coincident with an inertial frame S at time t = 0 and is moving with a constant velocity **v** relative to S in

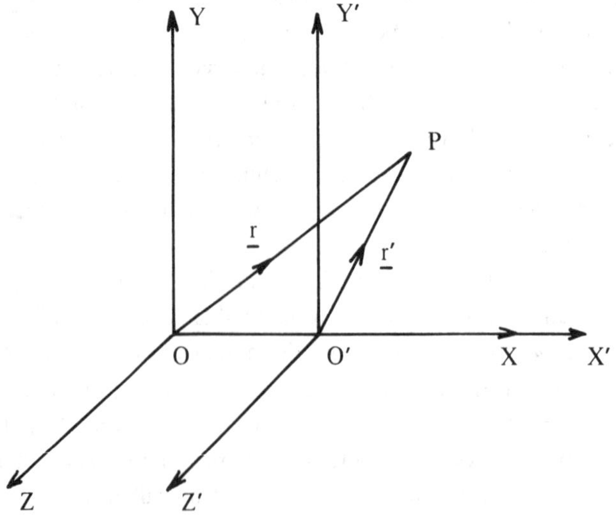

Fig. 1.1. The Galilean transformation

the positive direction of their common x-axes as shown in Fig. 1.1. We shall call S and S′ **standard frames** or **frames in standard configuration.** Let O′ be the position of the origin of S′ after a time t when an event occurs at a point P. Let **r** and **r**′ be the position vectors of the point P with respect to O and O′ respectively. Then from Fig. 1.1:

$$\overrightarrow{O'P} = \overrightarrow{OP} - \overrightarrow{OO'}$$

or $$\mathbf{r}' = \mathbf{r} - \mathbf{v}\, t. \tag{1.1}$$

Let x, y, z, and x′, y′, z′ be the coordinates of the same point P in the frames S and S′. Then

$$(x', y', z') = (x, y, z) - (v, 0, 0)\, t$$

so that

$$x' = x - v\, t, \tag{1.2a}$$

$$y' = y, \tag{1.2b}$$

$$z' = z. \tag{1.2c}$$

In classical physics, time is absolute so that it is unaffected by a motion of the frame of reference. Therefore if t′ denotes the time at which the event at P is observed from the frame S′, then

$$t' = t. \tag{1.2d}$$

That is, the observers in the frames S and S′ will assign the same time to any given event.

Equations (1.2) show how the space and time coordinates of an event in two standard frames of reference S and S′ are related to each other. This particular transformation of coordinates is called the **Galilean transformation** for frames in standard configuration. It is found to be valid experimentally whenever the relative velocity between S and S′ is small as compared to the velocity of light.

We are now well equipped to show that all frames of reference which are in uniform motion relative to an inertial frame are themselves inertial. To prove this, consider two standard frames of reference S and S′, such that S is an inertial frame. We have to show that S′ is also inertial.

According to the Galilean transformation, the position vectors of any particle in the two frames are related at any time t by equation (1.1). Differentiating this equation with respect to t, we get

$$\frac{d\mathbf{r}'}{dt} = \frac{d\mathbf{r}}{dt} - \mathbf{v}.$$

Since $t' = t$, we may write the above equation as

$$\frac{d\mathbf{r}'}{dt'} = \frac{d\mathbf{r}}{dt} - \mathbf{v}$$

or $$\mathbf{V}' = \mathbf{V} - \mathbf{v},$$

where $\mathbf{V}$ and $\mathbf{V}'$ denote respectively the velocities of the same particle as measured in the frames S and S′. Therefore if the velocity of a particle is constant in the frame S, then its velocity in S′ will also be constant. That is if S is an inertial frame, the frame S′ will also be inertial. This proves the above statement.

Covariance of Newton's Second Law of Motion

It is tempting to assume, and we do assume, that Newton's second law of motion, expressed symbolically by the vector equation

$$\mathbf{F} = \frac{d}{dt}(m\mathbf{V}) = m\,\frac{d^2\mathbf{r}}{dt^2}, \tag{1.3}$$

where $\mathbf{V}$ is the velocity of a particle of mass m and $\mathbf{F}$ is the force acting on it, is valid in an inertial frame. Clearly, for $\mathbf{F} = 0$, equation (1.3) is simply a symbolic expression for the first law.

We next show that the vector equation of motion has the same form in all inertial frames of reference.

Consider two standard frames of reference S and S′ related by the Galilean transformation (1.1). On differentiation with respect to t', equation (1.1) yields

$$\frac{d^2\mathbf{r}'}{dt'^2} = \frac{d^2\mathbf{r}}{dt^2},$$

where we have also used the fact that $t' = t$ and that the relative velocity $\mathbf{v}$ is constant. Since force and mass are taken as absolute quantities in Newtonian mechanics they should have the same values in all inertial frames of reference:

$$\mathbf{F}' = \mathbf{F} \quad \text{and} \quad m' = m.$$

Substituting the expressions for $\mathbf{F}$, m and $\frac{d^2\mathbf{r}}{dt^2}$ in equation (1.3), we get

$$\mathbf{F}' = m' \frac{d^2\mathbf{r}'}{dt'^2}.$$

This shows that the vector equation expressing Newton's second law of motion does not change in form in going from one inertial frame to another. This fact is expressed by stating that Newton's second law of motion or the corresponding equation is **covariant** under a Galilean transformation.

In general, if a law (or the equation expressing that law) does not change in going from one inertial frame of reference to another, it is said to be **covariant** under the transformation connecting the inertial frames. On the other hand, if a quantity or an expression remains unaltered in going from one inertial frame to another, it is said to be **invariant**. Thus in Newtonian mechanics, the force $\mathbf{F}$, the mass m and the time t are invariant quantities whereas the position $\mathbf{r}$ is not invariant as it changes to $\mathbf{r}'$ in going from one inertial frame to another. However, the equation

$$\mathbf{F} = m \frac{d^2\mathbf{r}}{dt^2}, \tag{1.3'}$$

does not change in such a transformation and is therefore covariant under the Galilean transformation connecting the inertial frames.

Newton's Second Law and Absolute Motion in Space

All the mechanical phenomena are governed by Newton's second law and the equation expressing this law has the same form in all inertial frames. This covariance has a great physical significance as identical mechanical experiments performed in any two inertial frames should yield identical results in both the frames and therefore it should not be possible to distinguish between different inertial frames by means of mechanical experiments. We would illustrate it by an example.

First of all, we note that if the rotation of the earth about its axis is ignored, then it may be regarded as an inertial frame because very distant objects which do not interact significantly with each other move almost in straight lines when observed from this frame. Then any frame of reference having a constant velocity relative to the earth is an inertial frame. Consequently, a train at rest relative to the ground ($\mathbf{v} = 0$) and a train moving with a uniform velocity relative to the ground are inertial frames. Consider a stationary train and suppose that a man in the train throws a ball vertically upwards. The subsequent motion of the ball is governed by Newton's second law according to which it would return to the point from where it was thrown up. Now suppose that the train starts moving and after some time attains a *constant velocity* relative to the ground. Since this train in uniform motion is also an inertial frame, Newton's second law would be valid in this frame too, and therefore a ball thrown vertically upwards in the train would return to the same point in the train from where it was thrown up. This means that if the man has to decide whether he is in uniform motion or at rest relative to the earth by arguing on the basis of throwing a ball vertically upwards he would never be able to do so because the ball would always return to the same point from which it is thrown up whether the train were in uniform motion or at rest relative to the ground. Thus this experiment cannot distinguish between two inertial frames of reference. In fact, it is impossible to design a mechanical experiment by which we can distinguish between different inertial frames of reference, because all mechanical phenomena are governed by Newton's second law which has the same form in all inertial frames.

Now we are well equipped to answer the all-important question: can we determine the speed of this train by performing a mechanical experiment inside the train? Certainly not. Because if we could, then we would be differentiating between two inertial frames of reference namely the train at rest and the train in uniform motion. And this is impossible owing to the covariance of Newton's fundamental equations of motion under a Galilean transformation. Therefore the speed of a uniformly moving train without reference to anything external to the train has no meaning at all as it cannot be determined by any mechanical experiment which is confined to the train alone. We can talk about the motion of the train only with respect to some external object, i.e., we can talk about the relative motion only. We therefore conclude that mechanical phenomena cannot determine absolute motion in space, i.e., the motion of an object without any reference to anything external. This is the principle of **Galilean relativity.**

It should be noted that the motion of the train relative to the ground can be detected by performing a mechanical experiment which is not confined to the train. For instance, suppose that two balls are allowed to fall, one after the other, to the ground from the same position, out of a window of the train. If the train is at rest the two balls would hit the same place on the ground. If, however, the train is in motion the two balls would hit the ground at different places indicating thereby a relative motion between the train and the ground. It may be emphasized that even this experiment tells us only about the relative motion of the train and the ground and not about the absolute motion of the train.

Maxwell's Equations and Absolute Motion in Space

We have seen that absolute motion in space cannot be determined by mechanical experiments. But is it possible to determine the absolute motion by an optical experiment? To answer this question, we have to consider the nature of light. Newton did not believe in the wave theory of light. But when Huygens established the wave nature of light, the problem of its propagation in vacuum became worth a careful consideration. Sound waves can travel only through a material medium, the particles of the medium vibrating about their mean positions constituting the wave. But if the waves are propagated only via material media, how is it possible for light travelling through vacuum to reach our planet from distant stars? This was explained by assuming the existence of a medium called **the ether** which pervaded all space. Later on, Maxwell showed that these light waves were electromagnetic in nature and their behaviour was governed by equations now known as **Maxwell's equations**. These equations are not covariant under a Galilean transformation. Therefore, in principle, optical experiments could be used to distinguish between different inertial frames of reference and hence to determine absolute motion in space.

Maxwell's equations in their well-known form are assumed to be valid in the ether which is supposed to represent the state of absolute rest. Thus if electromagnetic phenomena are observed in some inertial frame in motion relative to the ether the results would involve the velocity of that frame relative to the ether. Hence by an electromagnetic experiment, performed in an inertial frame without reference to any external object, we can detect absolute motion of the inertial frame, i.e., the velocity of the inertial frame relative to the ether. In particular, the absolute velocity of the earth can be determined by performing an optical experiment confined to the earth. An ingenious experiment for this purpose was designed and performed in 1887 by Michelson and Morley[1]. However, contrary to all expectations, the speed of the earth could not be detected by this experiment. This failure to determine absolute motion in space by an optical experiment is known as the **negative result of the Michelson-Morley experiment**. (See Appendix A.)

Before Einstein put forward his special theory of relativity, many unsuccessful attempts were made to explain the negative result of the Michelson-Morley experiment. Piecemeal explanations of various phenomena could be given but a unified explanation of the negative result of the Michelson-Morley experiment, the aberration of starlight and many other phenomena was not forthcoming. Of course, something was wrong somewhere.

2

BASIC POSTULATES OF SPECIAL RELATIVITY AND RELATIVISTIC KINEMATICS

Basic Postulates of Special Relativity

Einstein not only explained the negative result of the Michelson-Morley experiment, the aberration of starlight and other puzzling experiments but also opened up entirely new fields by advancing the following two fundamental postulates:

1. The laws of physics are the same in all inertial frames of reference.
2. The velocity of light in vacuum has the same constant value in all inertial frames and is independent of the motion of the light source.

The theory based on these postulates is called the **theory of the special relativity** or more frequently the **special theory of relativity**. It is said to be the special theory because it is valid only for a special class of frames namely the inertial frames. The first of these postulates means that the equations expressing a physical law should have the same form in all inertial frames so that it would not be possible to distinguish between these frames by any physical experiment. This is known as the **principle of relativity**. In fact, for classical mechanics this principle of relativity had been accepted ever since the days of Newton. Einstein merely extended its domain to the whole of physics. The second postulate appears to be against *common sense*. We know that the velocity of a material object, say that of a stone, thrown from a uniformly moving train, depends upon the speed of the train. Classically the same is true for the velocity of light; it depends upon the motion of its source. But, according to the second postulate, whatever the speed of the source of light, the velocity of light in vacuum is independent of the motion of the source; it always travels with the same speed. It must be emphasized that it is this principle of the *constancy of the velocity of light* which distinguishes the special theory of relativity from other theories, like Newtonian mechanics, which satisfy the principle of relativity.

The special theory of relativity immediately explains the Michelson-

10

Morley experiment, as it asserts that it is impossible to detect absolute motion in space.

In this chapter we will use postulates of special relativity to derive the transformations connecting the space and time coordinates of an event as measured in two inertial frames of reference and then examine the consequences of transformation.

The Lorentz Transformation

Consider two inertial frames S and S′, which are in standard configuration so that S′ is moving along the positive direction of the common axis of the two frames with a constant speed v relative to S (Fig. 2.1). Let the origins O and O′ of the two frames be coincident at time $t=t'=0$. Suppose that at that instant a point source fixed at the origin O of S emits a light pulse. This travels outward in all directions with the velocity c. Then at a later time, an experimenter in S will observe a spherical wave front with centre at O. At time t, the Cartesian equation of the spherical wave front of radius ct and centred at O will be

$$x^2 + y^2 + z^2 = c^2 t^2$$

or

$$x^2 + y^2 + z^2 - c^2 t^2 = 0. \tag{2.1}$$

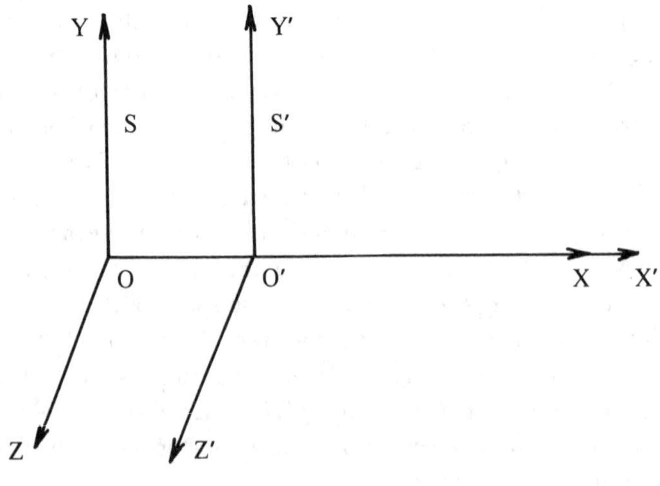

Fig. 2.1. Standard frames S and S′in relative motion

Since the speed of propagation of light is the same in all directions independently of the motion of the source, an observer in S' must also see a spherical wave front with centre at O' expanding with velocity c and therefore described by the equation

$$x'^2 + y'^2 + z'^2 - c^2 t'^2 = 0. \tag{2.2}$$

In order that there may be a 1–1 correspondence between the points of S and S', i.e., a single event in one inertial frame may transform to a single event in the other inertial frame, the connection between x', y', z', t' and x, y, z, t must be linear, i.e.,

$$x' = a_{11}x + a_{12}y + a_{13}z + a_{14}t + b_1$$

$$y' = a_{21}x + a_{22}y + a_{23}z + a_{24}t + b_2$$

$$z' = a_{31}x + a_{32}y + a_{33}z + a_{34}t + b_3$$

$$t' = a_{41}x + a_{42}y + a_{43}z + a_{44}t + b_4,$$

where a's may depend upon the speed v. The constant terms b_μ, $\mu = 1, 2, 3, 4$, on the right hand sides of these equations become zero because at $t = t' = 0$, the origins O and O' coincide. The above equations then reduce to

$$x' = a_{11}x + a_{12}y + a_{13}z + a_{14}t \tag{2.3a}$$

$$y' = a_{21}x + a_{22}y + a_{23}z + a_{24}t \tag{2.3b}$$

$$z' = a_{31}x + a_{32}y + a_{33}z + a_{34}t \tag{2.3c}$$

$$t' = a_{41}x + a_{42}y + a_{43}z + a_{44}t. \tag{2.3d}$$

If we substitute the expressions for x', y', z', t' from equations (2.3) in the left hand side of equation (2.2), it becomes a homogeneous quadratic expression, say Q(x, y, z, t). The equation $Q = 0$ must be satisfied by exactly the same sets of values x, y, z, t which satisfy equation (2.1). This is possible only if

$$x'^2 + y'^2 + z'^2 - c^2\,t'^2 = K(v)\,(x^2 + y^2 + z^2 - c^2\,t^2), \tag{2.4a}$$

where K may depend upon v but does not depend upon space and time coordinates. Since the frame S moves with a velocity $-v$ relative to S', we must have

$$x^2 + y^2 + z^2 - c^2\,t^2 = K(-v)\,(x'^2 + y'^2 + z'^2 - c^2\,t'^2). \tag{2.4b}$$

Comparing equations (2.4a) and (2.4b), we obtain

$$K(v)\,K(-v) = 1. \tag{2.5}$$

Assuming that space is isotropic, i.e., it possesses the same properties in all directions, K can depend only on the magnitude of velocity and not on its direction:

$$K(-v) = K(v).$$

By virtue of this relation, equation (2.5) reduces to $K^2 = 1$ so that $K = \pm 1$. For $K = -1$, equation (2.4a) becomes

$$x'^2 + y'^2 + z'^2 - c^2\,t'^2 = -\,(x^2 + y^2 + z^2 - c^2\,t^2). \tag{2.6}$$

For $v = 0$, the two frames of reference remain coincident at all times. In particular, for $t = 0$, the points on x-axis must satisfy the relation $x'^2 = x^2$ while for such points equation (2.6) gives

$$x'^2 = -\,x^2.$$

Since the two results are not consistent, the value $K = -1$ must be discarded. We, therefore, take $K = +1$ and equation (2.4a) thus reduces to

$$x'^2 + y'^2 + z'^2 - c^2\,t'^2 = x^2 + y^2 + z^2 - c^2\,t^2. \tag{2.7}$$

Although there is *a priori* no justification for the same, we assume that the planes $y = 0$ and $y' = 0$ of the two frames S and S' are permanently coincident. This assumption is justified by the results deduced from the relativistic transformation equations. Therefore if we put $y = 0$ in the right hand side of equation

$$y' = a_{21}(v)\,x + a_{22}(v)\,y + a_{23}(v)\,z + a_{24}(v)\,t, \tag{2.3b$'$}$$

we must obtain $y' = 0$ for all x, z and t. (*A prime after an equation number indicates that the corresponding equation occurred earlier.*) This is possible only if $a_{21} = a_{23} = a_{24} = 0$. Equation (2.3b), therefore, reduces to

$$y' = a_{22}(v)\,y. \tag{2.8}$$

As before, we must also have

$$y = a_{22}(-v)\,y'. \tag{2.9}$$

By virtue of relations (2.8) and (2.9), we obtain

$$a_{22}(v)\,a_{22}(-v) = 1.$$

But the isotropy of space demands that $a_{22}(v) = a_{22}(-v)$ and hence the above relation yields

$$a_{22}(v) = \pm 1.$$

Since at $t = 0$, (i) the origins of S and S$'$ are coincident and (ii) the y- and y$'$-axes coincide, the y- and y$'$-coordinates of any event are the same. Therefore, relations (2.8) and (2.9) can be valid only if $a_{22} = \pm 1$. Hence

$$y' = y.$$

Similarly, we can show that

$$z' = z.$$

Substituting these expressions for y and z in equation (2.7), we get

$$x'^2 - c^2 t'^2 = x^2 - c^2 t^2. \tag{2.10}$$

Now as the planes $x = 0$ and $x' = 0$ of S and S′ respectively are coincident at $t = t' = 0$, the vanishing of x and t together would imply the vanishing of x' and t' for all y and z. Equations (2.3a, d), therefore, give $a_{12} = a_{13} = a_{42} = a_{43} = 0$ and reduce to

$$x' = a_{11}\, x + a_{14}\, t \tag{2.11}$$

$$t' = a_{41}\, x + a_{44}\, t. \tag{2.12}$$

Now consider the origin O′ of the frame S′. Its position coordinate x' with respect to S′ is always zero: $x' = 0$ for O′, while its position coordinate x at a time t as measured from S is given by $x = v\,t$. Substituting these expressions in equation (2.11), we get

$$a_{14} = -\, a_{11}\, v.$$

Equation (2.11) then reduces to

$$x' = a_{11}\,(x - v\,t). \tag{2.13}$$

Next consider the origin O of the frame S. The position coordinate x of this point with respect to S is always zero, i.e., at the origin, $x = 0$ for all t, while at any time t' the position coordinate of this point with respect to S′ is given by $x' = -\,v\,t'$, because S is moving relative to S′ with a velocity $-v$. Substituting these expressions in equations (2.12) and (2.13), we get

$$t' = a_{44}\, t$$

and $$t' = a_{11}\, t,$$

respectively. These equations show that $a_{44} = a_{11}$, so that equation (2.12) reduces to

$$t' = a_{41}\, x + a_{11}\, t. \tag{2.14}$$

Substituting the expressions for x' and t' from equations (2.13) and (2.14) in equation (2.10), we get

$$a_{11}^2 (x^2 - 2\,v\,x\,t + v^2 t^2) - c^2 (a_{41}^2 x^2 + 2\,a_{41}\,a_{11}\,x\,t + a_{11}^2 t^2)$$
$$= x^2 - c^2 t^2. \qquad (2.15)$$

Since this equation must hold for all values of x, t, the coefficients of different powers of x and t on the two sides of the equation must be equal. Comparing the coefficients of t^2, we get

$$a_{11}^2 v^2 - c^2 a_{11}^2 = - c^2$$

or
$$a_{11} = \frac{1}{\sqrt{1 - v^2/c^2}}. \qquad (2.16a)$$

While taking the square root, only the positive sign is retained because $x' = a_{11}(x - v\,t)$ should reduce to $x' = x$ for $v = 0$. Comparing the coefficients of xt in equation (2.15), we get

$$- 2\,v\,a_{11}^2 - 2\,c^2\,a_{41}\,a_{11} = 0$$

which, on using relation (2.16a), gives

$$a_{41} = - \frac{1}{\sqrt{1 - v^2/c^2}} \frac{v}{c^2}. \qquad (2.16b)$$

Substituting the expressions for a and e from equations (2.16a,b) in equations (2.13) and (2.14), we get

$$x' = \frac{1}{\sqrt{1 - v^2/c^2}} (x - v\,t)$$

$$t' = \frac{1}{\sqrt{1 - v^2/c^2}} (t - \frac{v}{c^2} x).$$

It is customary to denote the expression $1/(1 - v^2/c^2)^{1/2}$ by $\gamma(v)$ or more frequently by γ. Then equations (2.3) assume the following form:

$$x' = \gamma\,(x - v\,t) \qquad (2.17a)$$

$$y' = y \qquad (2.17b)$$

$$z' = z \tag{2.17c}$$

$$t' = \gamma\left(t - \frac{v}{c^2}x\right). \tag{2.17d}$$

These are called the **Lorentz transformation equations**. They show how the coordinates of an event in two inertial frames of reference are related to each other.

Thus if two observers, one in S and the other in S$'$, and equipped with their own measuring sticks and clocks, make measurements of space and time coordinates of the same event, then their measurements would be related by the above equation. In view of the special configuration of the coordinate frames, the transformation is known as **special** or **standard Lorentz transformation** and replaces the Galilean transformation. In further discussion, unless explicitly stated otherwise, the *Lorentz transformation* would mean the *special Lorentz transformation*. It may be noticed that for small values of v/c, it is approximated by the Galilean transformation. This fact is also expressed by stating that if the velocity of light were infinite ($c \to \infty$), the Lorentz transformation would reduce to the Galilean transformation.

We remark that the Lorentz transformations serve as a dictionary in going from one inertial frame to another. Whenever we have to *translate* the space and time coordinates of an event as measured in one inertial frame to another, we make use of this dictionary.

Equation (2.17d) shows that the time scale is also transformed in going from one inertial frame of reference to another. Time is thus no more a universal parameter. The theory of special relativity changes it from an absolute quantity to one possessing relative character. In fact, the time at which an event occurs in the frame S$'$ depends not only upon the time at which that event occurs in S but also on the space coordinates of the point where the event occurs in S.

Let us next suppose that as observed in a frame S, two bullets strike the points (x_1, y_1, z_1) and (x_2, y_2, z_2) simultaneously at time t. Then the times t_1' and t_2' at which they strike those points as observed from S$'$ are given by

$$t_1' = \gamma\left(t - \frac{v}{c^2}x_1\right)$$

$$t_2' = \gamma\left(t - \frac{v}{c^2}x_2\right)$$

or $$t_2' - t_1' = \gamma\frac{v}{c^2}(x_1 - x_2).$$

This shows that if $x_1 \neq x_2$, then the times t_1' and t_2' are different so that the two events are not simultaneous for an observer in S′. If, however, $x_1 = x_2$, then $t_1' = t_2'$ and the events are also simultaneous for the observer in S′. Simultaneity is, therefore, no longer an absolute quality but is relative to the observer. This is called **the relativity of simultaneity.**

Solving equations (2.17), we get the inverse Lorentz transformation:

$$
\begin{aligned}
x &= \gamma\,(x' + v\,t') \\
y &= y' \\
z &= z' \\
t &= \gamma\,(t' + \frac{v}{c^2}\,x')
\end{aligned}
\qquad (2.18)
$$

These equations give the space and time coordinates of an event as measured in S in terms of space and time coordinates of the same event as measured in S′.

The Lorentz transformation is completely symmetric in the sense that inverse transformation can be obtained from the Lorentz transformation by merely interchanging the corresponding primed and unprimed quantities and replacing v by −v. This symmetry implies that none of the observers in the two inertial frames has a preferred viewpoint.

Problems

Derive relations (2.18).

Show that the Lorentz transformation leaves the expression $x^2 + y^2 + z^2 - c^2\,t^2$ invariant.

The Lorentz transformation equations show that the velocity of a material object would be always less than c. This can be seen quite easily. Suppose that a material object is moving along the x-axis of an inertial frame S. Then the object can be considered as instantaneous inertial frame of reference. If it is moving with a velocity $v > c$ with respect to S, the Lorentz transformation would lead to imaginary values of space and time coordinates.

This is impossible. Hence the velocity of any object cannot exceed the speed of light. For $v = c$, the denominators in the expressions for space and time coordinates become zero making the expressions indeterminate. Hence, in special relativity, the velocity of light is not only a universal constant but is also the upper bound of all physical velocities.

However, it must be stressed that the special relativity does not prevent us from considering geometrical velocities exceeding the velocity of light; it only asserts that information cannot be transmitted with a velocity greater than that of light. Thus if an observer stationed in a frame observes two objects moving in opposite directions, each with a velocity 0.8c, he will find that the two objects are separating from each other with a velocity 1.6c.

We shall now introduce the important geometrical concept of a 4-dimensional space-time continuum which was proposed by Minkowski in 1908.

We have seen that if x, y, z, t and x', y', z', t' are the space and time coordinates of the same event as noted by the observers in the two inertial frames of reference, then

$$x'^2 + y'^2 + z'^2 - c^2 t'^2 = x^2 + y^2 + z^2 - c^2 t^2.$$

Let us write x_1, x_2, x_3 and x_4 for x, y, z and ict, respectively, where $i = \sqrt{-1}$, and use a similar notation for the primed system. Then the above equation takes the form

$$x_1'^2 + x_2'^2 + x_3'^2 + x_4'^2 = x_1^2 + x_2^2 + x_3^2 + x_4^2.$$

This equation shows that (x_1, x_2, x_3, x_4) and (x_1', x_2', x_3', x_4') may be regarded as the coordinates of the same point in a 4-dimensional space in two orthogonal Cartesian coordinate frames with a common origin; each side representing the square of the distance of this point from the origin.

Since x_1, x_2, x_3 and $x_4 = ict$ locate an event in space and time, the point (x_1, x_2, x_3, x_4) may be considered as representing an event. The 4-dimensional space in which a point represents an event occurring at a certain point in ordinary space and at a certain time is called **Minkowski's space**. This space is also known as the **world space** or **space-time continuum.** Points and curves in this space are called the **world points** and the **world lines** respectively. It should be emphasized that Minkowski's space is not a straight generalization of the ordinary Euclidean 3-space to four dimensions. Owing to the negative sign of the coefficient of t^2 in the expression $x^2 + y^2 + z^2 -$

$c^2 t^2$, the time coordinate is not on the same footing as the space coordinates. This non-isotropic world space is actually pseudo-Euclidean in character. It is not possible to visualize the 4-dimensional space-time continuum, although, as we have just seen, it can be realized mathematically. However, we can draw 2-dimensional diagrams with one space and one time axis. We will develop relativistic dynamics by using the 4-dimensional formalism because it is not only simpler and more elegant than the conventional method, but also unifies concepts which were previously thought to be independent. Minkowski was so much overwhelmed by the beauty of the space-time world that he remarked, "*Space by itself and time by itself are to fade away into mere shadows and only a kind of union of the two will survive*". This prophecy has not come true. Special relativity does give a unifying picture of the various concepts, but time has never lost its individuality.

Writing x_1, x_2, x_3, x_4 for x, y, z, ict respectively in the Lorentz transformation equations (2.3), we get

$$x_1' = \gamma\,(x_1 + i\frac{v}{c}\,x_4)$$

$$x_2' = x_2$$

$$x_3' = x_3$$

$$x_4' = \gamma\,(x_4 - i\frac{v}{c}\,x_1).$$

These equations may be written in matrix form as $X' = AX$,

where $$X' = \begin{pmatrix} x_1' \\ x_2' \\ x_3' \\ x_4' \end{pmatrix}, \qquad X = \begin{pmatrix} x_1 \\ x_2 \\ x_3 \\ x_4 \end{pmatrix}$$

and

$$A = [a_{\mu\nu}] = \begin{pmatrix} \gamma & 0 & 0 & i\frac{v}{c}\gamma \\ 0 & 1 & 0 & 0 \\ 0 & 0 & 1 & 0 \\ -i\frac{v}{c}\gamma & 0 & 0 & \gamma \end{pmatrix}. \qquad (2.19)$$

The matrix A is called the **transformation matrix** and its determinant is equal to 1.

The Lorentz transformation equations can be written in another form by using a parameter α defined by

$$\tanh \alpha = \frac{v}{c}.$$

Problem

By using a parameter α defined by $\tanh \alpha = \frac{v}{c}$, show that the Lorentz transformation equations may be written as

$$x' = x \cosh \alpha - c\, t \sinh \alpha$$

$$x' = y$$

$$z' = z$$

$$c\, t' = c\, t \cosh \alpha - x \sinh \alpha.$$

Also show that the inverse transformation equations are given by

$$x = x' \cosh \alpha + c\, t' \sinh \alpha$$

$$y = y'$$

$$z = z'$$

$$c\, t = c\, t' \cosh \alpha + x' \sinh \alpha$$

Problem

By defining a parameter ϕ by $\tan \phi = i \frac{v}{c}$, show that the Lorentz transformation equations can be written as

$$x' = x \cos \phi + ict \sin \phi$$

$$y' = y$$

$$z' = z$$

$$ict' = ict \cos \phi - x \sin \phi.$$

Write down the inverse transformation equations.

Problem

Show that the Lorentz transformation in terms of spherical polar coordinates with the direction of relative motion as axis is given by

$$r' \cos \theta' = \gamma (r \cos \theta - v t)$$

$$r' \sin \theta' = \gamma \sin \theta$$

$$\phi' = \phi$$

$$t' = \gamma (t - \frac{v}{c^2} r \cos \theta).$$

The mathematical complexity of the Lorentz transformation is increased if the frames S and S′ are not in standard configuration although there is no increase in the kinematic information. We shall therefore restrict ourselves to the special Lorentz transformation. However, we shall make some comments and derive various types of Lorentz transformations.

(1) The special Lorentz transformation has been derived by assuming that the origins O and O′ of the two frames S and S′ are coincident at the time $t = t' = 0$, and that the uniform relative velocity between the two frames is along the positive direction of the common x-axis. This transformation is also called the **restricted Lorentz transformation.**

(2) Consider again two frames of reference in relative motion such that their respective axes are parallel to each other and their relative velocity is parallel to, say, their x-axes. Let us now rotate the coordinate axes in these frames through the same angle as measured in their own frames so as to obtain the coordinate systems S and S′. The relative velocity between these two frames will not be along the x- or x′-axis; the frame S′ would actually be moving with respect to S with a velocity $\mathbf{v} = (v_x, v_y, v_z)$. The corresponding Lorentz transformation is called the **general Lorentz transformation without rotation** of the coordinate axes. It means that by rotating back the coordinate

axes through the same angle as measured in their respective frames S and S′, we can bring the x-axes of both the frames in the direction of the relative velocity of the frames.

In order to obtain the general Lorentz transformation without rotation, consider the vector $\mathbf{r} = (x, y, z)$ drawn from the origin (0, 0, 0) of the frame S to the position (x, y, z) of an event occurring at a time t. Resolve this vector in two components $\mathbf{r}_L$ and $\mathbf{r}_T$, one parallel and the other perpendicular to the velocity $\mathbf{v}$:

$$\mathbf{r} = \mathbf{r}_L + \mathbf{r}_T,$$

where $$\mathbf{r}_L = (\mathbf{r} \cdot \mathbf{v})\,\mathbf{v}/v^2$$

and $$\mathbf{r}_T = \mathbf{r} - (\mathbf{r} \cdot \mathbf{v})\,\mathbf{v}/v^2.$$

Similarly, if $\mathbf{r}' = (x', y', z')$ and t' represent the position vector and time of occurrence of the same event in the frame S′, we may write $\mathbf{r}' = \mathbf{r}'_L + \mathbf{r}'_T$. We apply the special Lorentz transformation to the two components $\mathbf{r}_L$ and $\mathbf{r}_T$ separately and then combine them to form the vector $\mathbf{r}$. Since $\mathbf{r}_L$ and $\mathbf{r}_T$ are respectively along and perpendicular to the velocity $\mathbf{v}$ like x- and y- (or z-) coordinates in the standard configuration of frames, it follows that

$$\mathbf{r}'_L = \gamma\,(\mathbf{r}_L - \mathbf{v}\,t)$$

and $$\mathbf{r}'_T = \mathbf{r}_T.$$

Thus $$\mathbf{r}' = \mathbf{r}'_L + \mathbf{r}'_T = \gamma\,(\mathbf{r}_L - \mathbf{v}\,t) + \mathbf{r}_T$$

$$= \gamma\,[(\mathbf{r} \cdot \mathbf{v})\,\mathbf{v}/v^2 - \mathbf{v}\,t] + \mathbf{r} - (\mathbf{r} \cdot \mathbf{v})\,\mathbf{v}/v^2$$

$$= \mathbf{r} + \mathbf{v}\,[(\gamma - 1)\,(\mathbf{r} \cdot \mathbf{v})/v^2 - \gamma\,t]. \tag{2.20}$$

The transformation equation for time is given by

$$t' = \gamma\,[t - (\mathbf{r}_L \cdot \mathbf{v})/c^2]$$

$$= \gamma\,[t - (\mathbf{r} \cdot \mathbf{v})/c^2],$$

because the transverse component of **r** is perpendicular to **v**.

The inverse transformation equations are obtained by interchanging the primed and unprimed coordinates and replacing **v** by −**v**. We thereby get

$$\mathbf{r} = \mathbf{r}' + \mathbf{v}\,[(\gamma - 1)(\mathbf{r}'.\,\mathbf{v})/v^2 + \gamma\, t']$$

$$t = \gamma\,[t' + (\mathbf{r}'.\,\mathbf{v})/c^2].$$

(3) Let us next consider the case when the coordinate axes in S and S′ do not have the same orientation, i.e., when, say, the x- and x′-coordinate axes must be rotated through different angles as measured in their own frames in order to bring their x-axes in the direction of the relative velocity **v**. To obtain the formula for the Lorentz transformation we need only consider a further transformation through a rotation R of one of the frames, say S. In the matrix notation, this rotation, represented by the matrix R, will change the vectors **r** and **v** into (R**r**) and (R**v**) but the scalar product **r** • **v** will remain unchanged:

$$(R\mathbf{r}) \cdot (R\mathbf{v}) = \mathbf{r} \cdot \mathbf{v}.$$

Consequently, in this case the equations for the Lorentz transformation become

$$\mathbf{r}' = R\mathbf{r} + R\mathbf{v}\,[(\gamma - 1)(\mathbf{r} \cdot \mathbf{v})/v^2 - \gamma\, t]$$

$$t' = \gamma\,[t - (\mathbf{r} \cdot \mathbf{v})/c^2].$$

This transformation is known as the **general Lorentz transformation with rotation.**

(4) So far we have assumed that the origins O and O′ coincide at the time $t = t' = 0$, so that the Lorentz transformations are homogeneous transformations of the space-time coordinates. We shall now consider the case when the origins of the two frames S and S′ are not coincident and $t' \neq 0$ at $t = 0$.

Let S and S′ be two frames of reference in standard configuration but such that their origins are not coincident and $t' \neq 0$ at $t = 0$. Suppose that, at $t = 0$, a light signal is flashed from the origin O(0, 0, 0) of S. Let the same signal be observed to be emitted at t'_0 from the point (x'_0, y'_0, z'_0) in S′. Then the light wave front in S and S′ is described respectively by

$$x^2 + y^2 + z^2 - c^2 t^2 = 0$$

and $$(x' - x_0')^2 + (y' - y_0')^2 + (z' - z_0')^2 - c^2(t' - t_0')^2 = 0.$$

Thus the transformation equations must be such that

$$(x' - x_0')^2 + (y' - y_0')^2 + (z' - z_0')^2 - c^2(t' - t_0')^2 = x^2 + y^2 + z^2 - c^2 t^2.$$

Consequently, the Lorentz transformation corresponding to the motion of S′ relative to S along the common x-axis with constant velocity v is given by

$$x' - x_0' = \gamma\,(x - v\,t)$$

$$y' - y_0' = y$$

$$z' - z_0' = z$$

$$t' - t_0' = \gamma\,(t - \frac{v}{c^2}\,x).$$

Such a transformation is called **inhomogeneous Lorentz transformation.** The general inhomogeneous Lorentz transformation without rotation takes the form

$$\mathbf{r}' - \mathbf{r}_0' = \mathbf{r} + \mathbf{v}\,\{(\gamma - 1)\,(\mathbf{r} \cdot \mathbf{v})/v^2 - \gamma\,t\}$$

$$t' - t_0' = \gamma\,[t - (\mathbf{r} \cdot \mathbf{v})/c^2],$$

where $\mathbf{r}_0' = (\,x_0', y_0', z_0'\,)$.

The transformation equations for the **general inhomogeneous Lorentz transformation with rotation** are

$$\mathbf{r}' - \mathbf{r}_0' = R\mathbf{r} + R\mathbf{v}\,[(\gamma - 1)\,(\mathbf{r} \cdot \mathbf{v})/v^2 - \gamma\,t]$$

$$t' - t_0' = \gamma\,[t - (\mathbf{r} \cdot \mathbf{v})/c^2].$$

We will now show that two vectors which are perpendicular to each other in the frame S′ are in general not perpendicular to each other in the frame S. We have seen that

$$\mathbf{r}' = \mathbf{r} + \mathbf{v}\,[(\gamma - 1)\,(\mathbf{r} \cdot \mathbf{v})/v^2 - \gamma\, t]. \tag{2.20$'$}$$

Therefore if $\mathbf{r}_1'$ and $\mathbf{r}_2'$ are the position vectors of two fixed points in the frame S$'$, and $\mathbf{r}_1$ and $\mathbf{r}_2$ are the position vectors of the same points in the frame S, then, by virtue of equation (2.20$'$), we can write

$$\mathbf{r}_1' = \mathbf{r}_1 + \mathbf{v}\,[(\gamma - 1)\,(\mathbf{r}_1 \cdot \mathbf{v})/v^2 - \gamma\, t_1]$$

$$\mathbf{r}_2' = \mathbf{r}_2 + \mathbf{v}\,[(\gamma - 1)(\mathbf{r}_2 \cdot \mathbf{v})/v^2 - \gamma\, t_2].$$

Then the vector $\mathbf{x}' = \mathbf{r}_2' - \mathbf{r}_1'$ is given by

$$\mathbf{x}' = \mathbf{r}_2 - \mathbf{r}_1 + \mathbf{v}\,[(\gamma - 1)\,(\mathbf{r}_2 - \mathbf{r}_1)/v^2 \;- \gamma\,(t_2 - t_1)].$$

Since in the frame S, the positions of the two points must be measured simultaneously, we have $t_2 = t_1$. Then the above equation reduces to

$$\mathbf{x}' = \mathbf{x} + (\gamma - 1)\,(\mathbf{x} \cdot \mathbf{v})\,\mathbf{v}/v^2,$$

where $\quad \mathbf{x} = \mathbf{r}_2 - \mathbf{r}_1.$

The inverse transformation is obtained by interchanging $\mathbf{x}$ and $\mathbf{x}'$ and changing the sign of $\mathbf{v}$ so that

$$\mathbf{x} = \mathbf{x}' + (\gamma - 1)\,(\mathbf{x}' \cdot \mathbf{v})\,\mathbf{v}/v^2.$$

Now if $\mathbf{x}_1'$ and $\mathbf{x}_2'$ are any two fixed vectors in S$'$, then the corresponding vectors $\mathbf{x}_1$ and $\mathbf{x}_2$ in S are given by

$$\mathbf{x}_1 = \mathbf{x}_1' + (\gamma - 1)\,(\mathbf{x}_1' \cdot \mathbf{v})\,\mathbf{v}/v^2$$

$$\mathbf{x}_2 = \mathbf{x}_2' + (\gamma - 1)\,(\mathbf{x}_2' \cdot \mathbf{v})\,\mathbf{v}/v^2.$$

Multiplying $\mathbf{x}_1$ scalarly with $\mathbf{x}_2$, we obtain

$$\mathbf{x}_1 \cdot \mathbf{x}_2 = \mathbf{x}_1' \cdot \mathbf{x}_2' + 2\,(\gamma - 1)\,(\mathbf{x}_1' \cdot \mathbf{v})\,(\mathbf{x}_2' \cdot \mathbf{v})/v^2$$

$$+ (\gamma - 1)^2 (\mathbf{v} \cdot \mathbf{v}) (\mathbf{x}_1' \cdot \mathbf{v}) (\mathbf{x}_2' \cdot \mathbf{v})/v^4.$$

If the vectors $\mathbf{x}_1'$ and $\mathbf{x}_2'$ are perpendicular to each other, then $\mathbf{x}_1' \cdot \mathbf{x}_2' = 0$. However, $\mathbf{x}_1 \cdot \mathbf{x}_2$ will not be zero unless $\mathbf{x}_1' \cdot \mathbf{v}$ or $\mathbf{x}_2' \cdot \mathbf{v}$ is zero, i.e., unless one of the vectors $\mathbf{x}_1'$ and $\mathbf{x}_2'$ is perpendicular to $\mathbf{v}$. Thus, in general, two vectors which are perpendicular to each other in S′, will not be mutually perpendicular in S.

Problem

Show that if two fixed vectors are parallel in S′, they will not be parallel in S.

Special Lorentz Transformations as Rotations in Space-Time Continuum

A linear transformation $X' = A\,X + b$, where $A^T A = I$, is said to be an orthogonal transformation. This transformation relates the coordinates of a point in two rectangular coordinate systems inclined at a certain angle and relatively displaced. If the origins of the two frames are taken as coincident, the orthogonal transformation takes the form

$$X' = A\,X, \quad A^T A = I$$

and corresponds to a rotation.

In matrix notation, the special Lorentz transformations are described by

$$X' = A(v)\,X,$$

where

$$A(v) = \begin{pmatrix} \gamma(v) & 0 & 0 & -v\,\gamma(v) \\ 0 & 1 & 0 & 0 \\ 0 & 0 & 1 & 0 \\ -\gamma(v)\frac{v}{c^2} & 0 & 0 & \gamma(v) \end{pmatrix}, \quad \gamma = \frac{1}{\sqrt{1 - v^2/c^2}}\,.$$

It can be easily checked that $A^T(v)\ A(v) = I$. This shows that the special Lorentz transformation may be considered as an orthogonal transformation which represents a rotation in a 4-dimensional space-time continuum. Moreover det A = +1, where det A denotes the determinant of the matrix A. Such a Lorentz transformation is said to be a **proper orthogonal transformation.**

Consequences of the Lorentz Transformation

We shall now consider some of the consequences of the Lorentz transformation:

1. The Lorentz Contraction

Let us first study in detail the operation of measuring the length of an object in different frames of reference. Consider two standard inertial frames of reference S and S′, and a rod which is at rest along the x′-axis in the frame S′ (Fig. 2.2). Let ℓ be the length of this rod as measured by an observer in S′. Since the observer in S′ is at rest with respect to the rod, the length of the rod can be measured by noting the positions of the two ends of the rod, not necessarily at the same time. Let us see what would be the length of this rod as measured by an observer in S, i.e., by an observer who is at rest

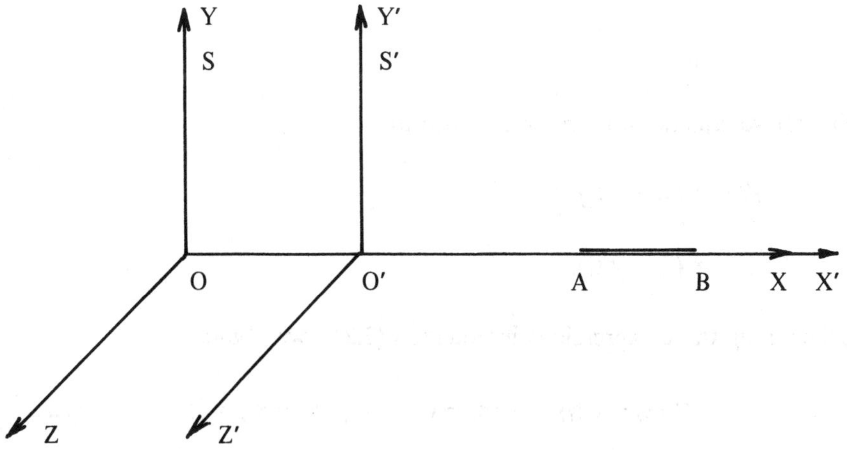

Fig. 2.2. The rod AB is stationary in S′

in S. Since the rod is in motion relative to the observer in S, the length of the rod can be measured correctly only by noting simultaneously the end positions of the rod. This direct matching of a swiftly moving rod and measuring stick is difficult in practice. However, we will assume that this can always be done. Let x_1, x_2 be the end positions of the rod measured simultaneously by an observer in S and x_1', x_2' be the positions of the end points of the same rod as measured by an observer in S′. Then the length ℓ of this rod as measured by the observer in S′ is given by

$$\ell' = x_2' - x_1' \, . \tag{2.21}$$

Now, by the Lorentz transformation, x_2' and x_1' are related to the space and time coordinates of the end points of the rod as measured in S by the equations

$$x_2' = \gamma\,(x_2 - v\,t_2)$$

$$x_1' = \gamma\,(x_1 - v\,t_1),$$

where t_1 and t_2 are the times at which the end positions x_1, x_2 of the rod are measured in the frame S. Since the coordinates x_1 and x_2 are measured simultaneously, we must have

$$t_2 = t_1.$$

The above equations, therefore, reduce to

$$x_2' = \gamma\,(x_2 - v\,t_1)$$

$$x_1' = \gamma\,(x_1 - v\,t_1).$$

Substituting these expressions in equation (2.21), we obtain

$$\ell' = \gamma\,(x_2 - v\,t_1) - \gamma\,(x_1 - v\,t_1) = \gamma\,(x_2 - x_1) = \gamma\,\ell, \tag{2.22}$$

where $\ell = x_2 - x_1$ is the length of the rod as measured in S. Since, the velocity v of a material object is always less than c, the factor $\gamma = 1/(1 - v^2/c^2)^{1/2}$ is always greater than unity. Thus from equation (2.22) we conclude that $\ell' > \ell$

or $\ell < \ell'$. This means that the length of the rod as measured by an observer who is in motion with respect to it is shorter and is reduced to $(1 - v^2/c^2)^{1/2}$ times its length as measured by an observer who is at rest with respect to it. In other words: for an observer, a rod in uniform motion is shortened in length. It may be noticed from the Lorentz transformation equations that if the rod is perpendicular to the direction of motion, then the length remains invariant. Thus contraction in the length takes place only in the direction of motion. This contraction in length of an object is called the **Lorentz-Fitzgerald contraction.** The length of the rod is greatest for the observer who is at rest relative to it. This length is called the **proper length** or the **rest length** of the rod. Since in the special relativity, information cannot be transmitted with a velocity greater than that of light, the above analysis is valid even when we are considering the distance between two points in space and not a material rod.

The phenomenon of length contraction is a reciprocal effect. That is if two rods having the same proper length are placed along the common x-axis such that one is at rest in S while the other is at rest in S′, the observer in each frame would find the rod in the other frame to be shorter by a factor $(1 - v^2/c^2)^{1/2}$. Thus if two persons carrying identical poles run past each other, each would observe the other's pole to be shorter than his own.

It has not been possible so far to directly verify the Lorentz contraction; one of the fundamental conclusions of the special relativity still awaits a direct confirmation. This is because of the non-availability of an object of appreciable length that moves with a speed comparable to that of light.

The Lorentz contraction can be measured, in principle, without any reference to clocks in the following way. Consider two identical rods AB and A′B′ moving in opposite directions at equal speeds, along a measuring stick PP′ which is at rest in a frame of reference S. Since the speeds of the rods

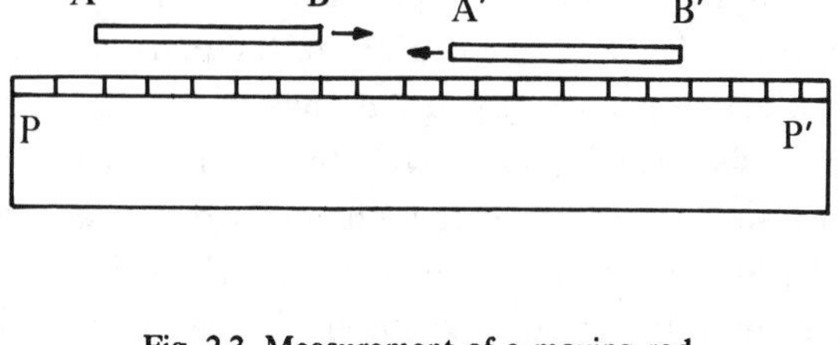

Fig. 2.3. Measurement of a moving rod

are equal, they must have equal lengths in S. Therefore, if at some instant, the left ends A, A′ coincide, then the right ends B, B′ would also coincide. Two observers can note the markings on PP′ where the coincidence of end points occurs. The distance between these markings gives the length of each one of the rods as observed in the frame S and would be less than the proper lengths of the rods. This process thus involves only the measurement of length. Practical difficulties, however, render this a hypothetical experiment.

Problem

The proper length of a rod is 10 m. (i) Calculate the length of the rod when it is moving with a velocity of (a) 3000 km s^{-1} (b) 0.99 c, where c is the velocity of light in vacuum. ($c = 3 \times 10^8$ m s^{-1}). (ii) Calculate the change in the length of the rod when it is moving with a velocity of 3 km/hour.

Is the Lorentz contraction *real* or *apparent*? Before answering this question, we note that, through generations, man has got this idea stuck in his mind that length is an absolute quantity. Accordingly, the length of an object should have the same value for all the observers whether in motion or at rest relative to it, and a change in length would mean that the length has either increased or decreased for all the observers. The theory of special relativity, however, asserts that the length of an object does not possess absolute significance. It depends upon the speed of the object relative to the frame of reference in which the measurement is made, and has, therefore, only a relative character. For this reason, the length of an object when in motion is not equal to its length when it is at rest. Since a measurement will certainly indicate this change, the Lorentz contraction is real.

Visual Appearance of Moving Objects

The measured length of a rod in a frame S is defined as the distance between positions of the ends of the rod, noted simultaneously if the rod is in motion relative to that frame. The positions of the ends at a time t are noted by observers who are in the vicinity of the ends of the rod at that time. The length of the rod obtained in this way is the measured length of the rod. However, when a single observer sees a moving rod or takes its photograph, its view at a particular time is determined by the light which, coming from the different parts of the rod, reaches the eye of the observer or the lens of the camera at that time. Obviously, the light which is scattered from the various parts of the rod at a certain time cannot reach an observer simultaneously, because at that time its different parts will be at different distances from the observer. Therefore the light which is received by the eye of an observer or the lens of a camera is not the light which leaves the various parts of the rod simultaneously. In fact, the rays of light which determine the view of the rod may leave the various parts of the rod at different times. The visual appearance of a moving rod would therefore be different from its measured length. Let us now see what will be the shape of a lamina as seen by an observer or by a camera.

Consider a square lamina ABCD of proper length ℓ. Let it be moving relative to an observer O with a velocity v as shown in Fig. 2.4a, i.e., moving in the xy-plane with the edge BC perpendicular to the y-axis. Let us suppose that the square is either of a small size or is far away from O so that the light scattered from the square reaches O, as a parallel beam. The sides BC and AD will be contracted to $\ell\,(1 - v^2/c^2)^{1/2}$, but each of the sides AB and DC, being perpendicular to v, will have the same measured length ℓ. Let us now see how the lamina will *appear* to the observer at O. When the rays of light from the various points of the side BC reach O at a particular instant, the rays originated a little earlier from other parts of the lamina can also reach O at the same time. For example, the ray of light emitted from A when it was at A′ will reach O at the same time as the ray from B reaches O, provided the time taken by A to travel from A′ to A is the same as that in which light covers a distance ℓ, the length of the edge AB, i.e.,

$$\frac{\ell}{c} = \frac{AA'}{v}$$

or $$AA' = \frac{\ell\,v}{c}\,.$$

This means that the rays of light coming from the edge AB convey the projection of A′B on BC to the observer at O. This projection is of length $\ell\,v/c$. Thus the edge of the lamina will appear to have length AA′ + BC = $\ell\,v/c + \ell\,(1 - v^2/c^2)^{1/2}$.

Now consider a similar square lamina ABCD at rest relative to O but rotated in the xy-plane through an angle α as shown in Fig. 2.4b. Then the projection of BD on a line parallel to the x-axis is $\ell \sin\alpha + \ell\cos\alpha$. If $\alpha = \sin^{-1} v/c$, then this projection is $\ell\,v/c + \ell\,(1 - v^2/c^2)^{1/2}$ which is the same as the length which the observer O sees in the case of a square lamina of side ℓ

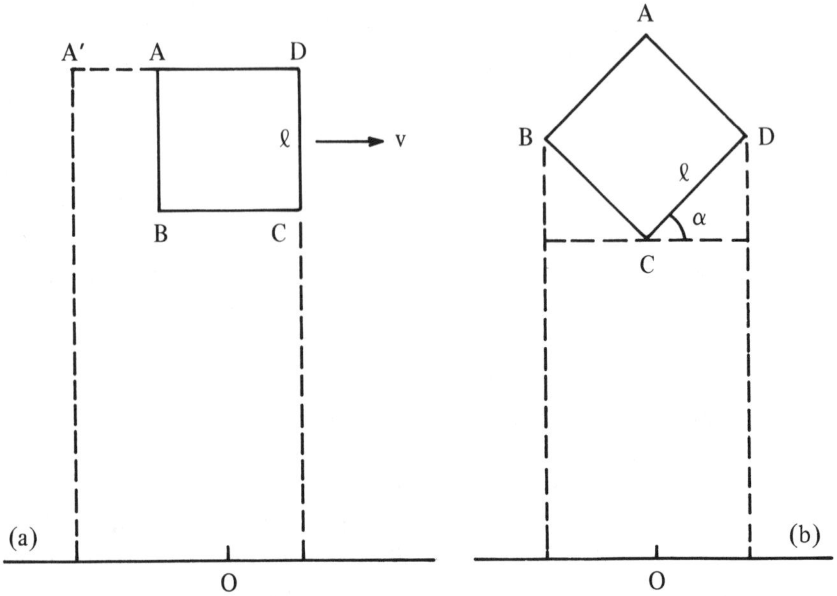

Fig. 2.4. Visual appearance of
(a) a rapidly moving square lamina and (b) a square lamina turned through an angle α

moving relative to him with velocity v. Therefore, as far as the visual appearance is concerned, to the observer O, a square lamina of proper side ℓ moving with a velocity v relative to him is the same as a stationary square lamina of side ℓ turned through an angle $\alpha = \sin^{-1} v/c$. Similarly, a rapidly moving cube, either of small size or far way from the observer, will appear the same way as a stationary cube turned through an angle which depends on the velocity of the cube.

Just like the square lamina, the view of a rapidly moving circular lamina is the same as that of a similar stationary lamina rotated in its plane through a certain angle depending upon the velocity of the moving lamina. Since a circle appears the same from all directions in its plane, a moving circle will offer the same view as the stationary circle; the effect of Lorentz contraction will not be visible. However, if measurements of different diameters of the circle are made, they will be different. Similarly, the view of a rapidly moving sphere will be the same as that of a similar stationary sphere.

Synchronization of Clocks

Before we take up further study of the consequences of the Lorentz

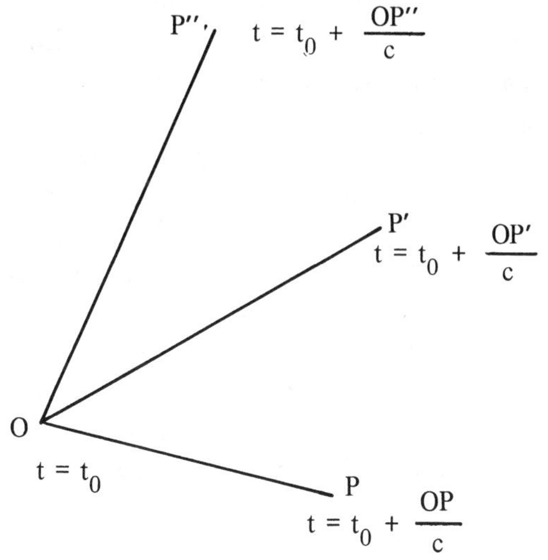

Fig. 2.5. Synchronization of clocks by light signal

transformation, it will be appropriate to consider the process of time measurement in detail. Let us first see how we can synchronize two clocks A and B stationed in a frame of reference. The easiest way to do this would be to bring one of these clocks, say A, in the vicinity of B, and set them to read the same. If we carry the clock A back to its original position, will the two clocks still agree? The answer to this question is that if the time interval noted on a clock does not depend upon the speed of the clock, then the two clocks will agree; otherwise they will be keeping different times. In order to synchronize distant clocks in a frame without any displacement from their positions, we may adopt the following method. We measure the distances of the various clocks, which are to be synchronized, from one of the clocks, say at the point O, and station an observer behind each clock. We inform all the observers that a flash of light will be emitted from a light source at O at time $t = t_0$ and that, as soon as they receive that flash, they should set their clocks located at points P, P′, P″, • • •, to read respectively $t_0 + OP/c$, $t_0 + OP'/c$, $t_0 + OP''/c$, • • •, where c is the velocity of light. This is shown in Fig. 2.5. Then these clocks would be synchronized with the clock at O and with each other. The clocks which have been regulated in this manner will remain synchronized provided they are not disturbed.

Can a single clock in a frame of reference be sufficient to measure the time of occurrence of any event? It would be so, provided that some signal from the event reaches the clock, and we know the velocity v of the signal as well as the distance r of the clock from the point where the event has occurred. If the signal reaches the clock at time t, then the event must have occurred at time $t - r/v$. However, a much more convenient process would be to have a clock in the vicinity of the event, and note the time directly on that clock. In principle, this would be possible if synchronized clocks are distributed throughout the frame of reference, and an observer is stationed near each clock. Then whenever an event occurs, the time would be recorded on the clock placed in its vicinity, by the observer stationed nearby. In further discussion the words *time noted on a clock in a frame* would mean the *time noted on a clock, placed in the vicinity of the event, by an observer (which may be a camera) stationed nearby.*

2. *Time Dilation*

Let us now study the effect of the Lorentz transformation on the measurement of time. Consider two events, say the emission of two sparks, occurring at times t_1' and t_2' in the frame S', where $t_2' > t_1'$. Then in this frame the time interval between the sparks is $t_2' - t_1'$. Denote it by T'. Let us suppose that an observer in S observes the same sparks at times t_1 and t_2. Denote the time interval $t_2 - t_1$ by T. Then, by using equations (2.19), we obtain

$$\begin{aligned} T &= t_2 - t_1 \\ &= \gamma\left(t_2' + \frac{v}{c^2} x_2'\right) - \gamma\left(t_1' + \frac{v}{c^2} x_1'\right) \\ &= \gamma\,(t_2' - t_1') + \gamma \frac{v}{c^2}\,(x_2' - x_1'). \end{aligned}$$

If in S' the two events take place at points for which x'-coordinates are equal, i.e., $x_2' = x_1'$, then the above equation reduces to

$$T = \gamma(t_2' - t_1') = \gamma\, T'. \tag{2.23}$$

Equation (2.23) shows that for the two events having the same x'-coordinates the time interval T as measured by an observer in S would be longer than the interval T' between these very events as measured by an observer in the frame S'. If the two events are the ticks of a clock C' in the frame S', the observer in S, after comparing the readings of this clock C′ with the readings of the synchronized clocks fixed in the frame S would find that the clock in S′ is going slow. Similarly, the observer in the frame S′ would find that the clock in S is going slow. This effect is called time-dilation and is usually expressed by saying that the moving clocks run slow as compared to the clocks at rest. If the moving clock is brought to rest, it regains its initial rate of flow of time. However, it will now lag behind in time when compared to the clocks which remained at rest. This delay in time was accumulated during the period of its motion, and unlike the Lorentz contraction, cannot be made good automatically.

Time dilation, like length contraction, is a real phenomenon. It may again be pointed out that the theory of relativity does not cast any new light on the nature of time; it only changes it from an absolute quantity to one possessing relative character.

Meson Decay

A very interesting example of time dilation is furnished by mu-mesons (μ-mesons), also called muons. These tiny particles are created at an altitude of more than 6000 m by cosmic rays coming from the outer space. These mesons have an average life of 2×10^{-6} s in their rest frame and travel towards the earth with a typical speed of 2.994×10^8 m s^{-1}. According to classical mechanics, before decaying, on an average they should travel a distance

$$d = \text{velocity of light} \times \text{mean life time of a } \mu\text{-meson}$$

$$= 2.994 \times 10^8 \times 2 \times 10^{-6} \text{ m}$$

$$= 600 \text{ m.}$$

Since mesons are created at altitudes which are about 10 times greater than this value, they should have virtually no chance to reach the sea level. Actually they reach the sea level in profusion. The abundance of these particles at sea level can be accounted for satisfactorily in terms of the theory of special relativity. We shall examine this problem from the inertial frame of an observer on the ground.

Let $\Delta\tau$ denote the average life time of a meson as measured in its rest frame and Δt denote its extended life time, i.e., the average time interval between the creation and disintegration of a meson, as measured from the frame of reference of an observer on the ground. Then as

$$\Delta\tau = 2 \times 10^{-6} \text{ s}$$

and $$v = 2.994 \times 10^{8} \text{ m s}^{-1},$$

we have $$\Delta t = \Delta\tau/(1 - v^2/c^2)^{1/2} = 31.7 \times 10^{-6} \text{ s}.$$

Thus, as noted by an observer on the ground, the average life of a meson, moving with a velocity of 2.994×10^{8} m s^{-1}, is 31.7×10^{-6} s, almost 16 times greater than when it is at rest. Therefore, during its life time it can travel a distance

$$v\,\Delta t = 2.994 \times 10^{8} \times 31.7 \times 10^{-6} \text{ m} = 9500 \text{ m}$$

and hence can reach the sea level before decaying.

This result establishes the fact, that life time is not an absolute quantity but has only a relative significance.

Problem

Solve the meson problem by examining it from the frame of reference of the meson.

The Twin Paradox

One of the most interesting and frequently discussed paradoxes in special relativity is the **twin paradox**. Consider twin brothers A and B and suppose that one of them, say A, decides to take a voyage into space. He starts his journey in a rocket, accelerates for a short time, attains a high velocity, moves with uniform velocity, reverses its direction after traversing a long distance, decelerates near the end of his journey, and ultimately lands on the earth. From the viewpoint of the stay-at-home twin B, the clocks in the rocket would be going slow, because most of the time the rocket had been moving with a uniform velocity relative to the earth. Thus from the standpoint of the twin B, when the travelling twin A returns to the earth, he should be younger than B. This difference in their ages would be appreciable if the rocket had travelled at a speed close to that of light. For instance, if, according to the earthbound twin B, the rocket had moved with a uniform speed 0.8c and had taken 25 years for the round trip, then according to the twin B, the time elapsed at the end of the journey would only be $25\,(1 - 0.8^2)^{1/2} = 15$ years. That is, at the end of the trip, the twin A who was travelling in the rocket would be 10 years younger than the stay-at-home twin B. The paradox arises when we consider the same problem from the point of view of the twin A, who, during his uniform motion, can regard himself at rest in the rocket and consider the earth moving away and coming back to him, so that on meeting B again he should find the twin B younger than himself. That is, at their rendezvous each one of the twins would find the other one younger. Such a situation is evidently absurd. This paradoxical situation is referred to as the **twin paradox**. It is also known as the **clock paradox** because instead of twins, we could compare two synchronized clocks, one of them remaining on the earth and the other taking a round trip in space with the rocket. In fact, there is no paradox at all. This problem has arisen because of a misunderstanding of the situation. During this round trip of A, twin B has all along been in one inertial frame. On the other hand, when the twin A starts the voyage, he must accelerate away. When he reverses the direction of the rocket, his frame of reference is accelerated relative to the earth. Moreover, the rocket accelerates while coming to rest with respect to the earth. Thus the positions of the twin A, who has been in an accelerated frame for a part of his journey and the twin B who has been in an inertial frame throughout this period are not symmetrical. This asymmetry arises because A has to change his inertial frames during the return trip. All observers will therefore agree that it is he who is in motion and that the clocks in the rocket are running slow. Thus there is no paradox. A genuine paradox would have arisen only if contradicting conclusions could be arrived at from symmetrical positions. If this asymmetrical situation is taken into consideration, the twin A returning from the distant star would be younger than the twin B who had stayed at home.

This can be seen as follows. The laws of special relativity are valid only for inertial frames of reference. Since the rocket in which the twin A starts, outward journey is accelerated for a very short time interval and then attains a uniform speed, he spends only a negligible time in an accelerated frame. In order to return to the earth, A must decelerate and reverse his motion. Although this also can be done in a very short time interval, the return journey of A occurs in a completely different inertial frame. As, while landing on the earth, the rocket can be brought to rest in a very short interval of time, it is the return journey of A in an inertial frame which is different from the inertial frame in which it was at rest on the outward journey, that causes the difference in the ages of the twins. As only A has to jump from one inertial frame to another to return, all observers will agree that it is he who is in motion and that the clocks in the rocket are running slow. Consequently, it is A who would be the younger twin on his return to the earth.

Transformation Law for Velocity

We shall now find the relativistic transformation law for velocity, i.e., we should like to see how the velocity in one inertial frame would appear from another inertial frame of reference. Let us consider the motion of a particle with respect to two standard frames of references S and S′. If the space-time coordinates of the particle in the two frames are (x, y, z, t) and (x′, y′, z′, t′), then its velocities $\mathbf{V} = (V_x, V_y, V_z)$ and $\mathbf{V}' = (V_x', V_y', V_z')$ as measured in these frames are given by

$$\mathbf{V} = \left(\frac{dx}{dt}, \frac{dy}{dt}, \frac{dz}{dt}\right)$$

$$\mathbf{V}' = \left(\frac{dx'}{dt'}, \frac{dy'}{dt'}, \frac{dz'}{dt'}\right).$$

We have proved earlier that

$$x = \gamma\,(x' + v\,t')$$

$$y = y'$$

$$z = z'$$

$$t = \gamma\,(t' + \frac{v}{c^2}\,x').$$

Differentiating with respect to t, we get

$$\frac{dx}{dt} = \gamma\left(\frac{dx'}{dt'} + v\right)\frac{dt'}{dt} \tag{2.24a}$$

$$\frac{dy}{dt} = \frac{dy'}{dt'}\,\frac{dt'}{dt} \tag{2.24b}$$

$$\frac{dz}{dt} = \frac{dz'}{dt'}\,\frac{dt'}{dt} \tag{2.24c}$$

$$1 = \gamma\left(1 + \frac{v}{c^2}\frac{dx'}{dt'}\right)\frac{dt'}{dt}. \tag{2.24d}$$

Substituting the expression for dt′/dt from equation (2.24d) in equation (2.24a), we get

$$\frac{dx}{dt} = \left(\frac{dx'}{dt'} + v\right)\Big/\left(1 + \frac{v}{c^2}\frac{dx'}{dt'}\right)$$

or

$$V_x = \frac{V'_x + v}{1 + \frac{v}{c^2}V'_x} \tag{2.25a}$$

Similarly, we can show that

$$V_y = \frac{V'_y}{\gamma\left(1 + \frac{v}{c^2}V'_x\right)} \tag{2.25b}$$

and

$$V_z = \frac{V'_z}{\gamma\left(1 + \frac{v}{c^2}V'_x\right)}. \tag{2.25c}$$

Equations (2.25) give the relativistic transformation laws for the components of velocity. Notice that in the case of velocity, in general, even the transverse components in the two frames differ from each other.

For very small values of v/c, equations (2.25) approximate to

$$V_x = V'_x + v \tag{2.26a}$$

$$V_y = V'_y \tag{2.26b}$$

$$V_z = V'_z, \tag{2:26c}$$

which is the classical transformation law for the velocity. Since the transformation equations (2.26) give the classical law of addition of velocities $\mathbf{V}' = (V'_x, V'_y, V'_z)$ and $\mathbf{v} = (v, 0, 0)$, equations (2.25) may be regarded as representing the relativistic law of addition of velocities $\mathbf{V}'$ and $\mathbf{v}$. However, ordinary mechanical velocities, being too small as compared with c, do not provide a suitable test of the relativistic law for the addition of velocities.

It is interesting to notice that the relativistic law of addition is such that if we add any velocity to the velocity of light it remains unchanged. For example, a light signal from a source at rest on the x'-axis in S' will be observed in S, in accordance with equation (2.25a), to be moving along the x-axis with a velocity

$$V_x = \frac{c + v}{1 + \frac{v}{c^2}\, c} = c.$$

Using equations (2.25), we obtain the following relation

$$c^2 - V^2 = \frac{c^2\,(c^2 - V'^2)(c^2 - v^2)}{(c^2 + V'\,v)^2}, \tag{2.27}$$

where $V^2 = V_x^2 + V_y^2 + V_z^2$ and $V'^2 = V_x'^2 + V_y'^2 + V_z'^2$.

For $V' < c$ and $v < c$, the right hand side and consequently the left hand side of equation (2.27) is positive, i.e.,

$$c^2 - V^2 > 0.$$

Therefore $V < c$. This shows that the resultant of two velocities V' and v, each of which is less than c, is itself less than c. Hence it is impossible to attain the velocity of light c by adding velocities each one of which is less than c.

The inverse transformation law for velocities is obtained from equations (2.26) by interchanging primed and unprimed quantities and changing v to $-v$, so that

$$V_x' = \frac{V_x - v}{1 - \frac{v}{c^2}\, V_x}, \tag{2.28a}$$

$$V_y' = \frac{V_y}{\gamma\,(1 - \frac{v}{c^2}\, V_x)}, \tag{2.28b}$$

$$V_z' = \frac{V_z}{\gamma\,(1 - \frac{v}{c^2}\, V_x)}. \tag{2.28c}$$

Problem

Show that the magnitude of the resultant 3-velocity $\mathbf{V}'$ is given by

$$V'^2 = [V^2 - 2\,\mathbf{V}\cdot\mathbf{v} + v^2 - (\mathbf{V}\wedge\mathbf{v})^2/c^2]/(1 - \mathbf{V}\cdot\mathbf{v}/c^2).$$

Rapidity or Pseudovelocity

According to the special theory of relativity, the resultant w of two *collinear* velocities u and v is given by

$$w = \frac{u + v}{1 + \frac{v}{c^2}u}. \tag{2.29}$$

This shows that in special relativity, in contrast to the classical law for the addition of collinear velocities, the velocities cannot be added algebraically. We can, however, define a quantity called **rapidity** or **pseudovelocity** which is related to the velocity and obeys the classical law for the addition of collinear velocities. We define the rapidity y_V associated with the velocity V by the equation

$$y_V = c \tanh^{-1}\frac{V}{c},$$

Then

$$V = c \tanh\frac{y_V}{c}.$$

Expressing the collinear velocities u, v, w in terms of the associated rapidities and substituting the expressions in equation (2.29), we get

$$\tanh\frac{y_w}{c} = \left(\tanh\frac{y_u}{c} + \tanh\frac{y_v}{c}\right)/\left(1 + \tanh\frac{y_u}{c}\tanh\frac{y_v}{c}\right)$$

$$= \tanh\left(\frac{y_u}{c} + \frac{y_v}{c}\right)$$

or

$$y_w = y_u + y_v,$$

i.e., the rapidities associated with the collinear velocities are added algebraically.

Problem

Find the transformation law for the rapidity.

Transformation Law for Acceleration

Let us next find the transformation law for acceleration. This may be obtained from the transformation law for velocity by differentiation. Thus differentiating equation (2.28) with respect to t′, we get

$$\frac{dV_x'}{dt'} = \left[\left(1 - V_x \frac{v}{c^2}\right)\frac{dV_x}{dt} - (V_x - v)\left(-\frac{v}{c^2}\frac{dV_x}{dt}\right)\right]\frac{dt}{dt'}$$

or

$$a_x' = \frac{1 - \frac{v^2}{c^2}}{\left(1 - V_x \frac{v}{c^2}\right)^2}\left(a_x \frac{dt}{dt'}\right),$$

where

$$a_x' = \frac{dV_x'}{dt'} \quad \text{and} \quad a_x = \frac{dV_x}{dt}. \tag{2.30}$$

But

$$t' = \gamma\left(t - \frac{v}{c^2}x\right),$$

which on differentiation with respect to t yields

$$\frac{dt'}{dt} = \gamma\left(1 - \frac{v}{c^2}\frac{dx}{dt}\right) = \gamma\left(1 - \frac{v}{c^2}V_x\right).$$

Substituting this expression in equation (2.30) and simplifying, we obtain

$$a_x' = \frac{\left(1 - \frac{v^2}{c^2}\right)^{3/2}}{\left(1 - V_x \frac{v}{c^2}\right)^3} a_x . \tag{2.31a}$$

Similarly, we can show that

$$a_y' = \frac{1 - \frac{v^2}{c^2}}{\left(1 - V_x \frac{v}{c^2}\right)^2}\left[a_y + \frac{V_y \frac{v}{c^2} a_x}{1 - V_x \frac{v}{c^2}}\right] \tag{2.31b}$$

and

$$a_z' = \frac{1 - \frac{v^2}{c^2}}{\left(1 - V_x \frac{v}{c^2}\right)^2}\left[a_z + \frac{V_z \frac{v}{c^2} a_x}{1 - V_x \frac{v}{c^2}}\right] \tag{2.31c}$$

Equations (2.31) give the transformation law for acceleration.

4-Vectors

A 4-vector A_μ in Minkowski's space-time continuum is defined as a quantity having four components A_1, A_2, A_3, A_4 which transform under a Lorentz transformation in the same way as space-time coordinates x_1, x_2, x_3, x_4 = ict, so that

$$A_1' = \gamma\,(A_1 + i\frac{v}{c}A_4)$$

$$A_2' = A_2$$

$$A_3' = A_3$$

$$A_4' = \gamma\,(A_4 - i\frac{v}{c}A_1)\,.$$

For a given observer the first three components x_1, x_2, x_3 of x_μ behave like an ordinary 3-vector while the fourth component behaves as a scalar in 3-dimensional space. Since under a Lorentz transformation every 4-vector behaves the same way as x_μ, this result is true for every 4-vector.

Squaring and adding the above four equations, we get

$$A_1'^2 + A_2'^2 + A_3'^2 + A_4'^2 = A_1^2 + A_2^2 + A_3^2 + A_4^2\,.$$

This result shows that the length of a 4-vector, viz., $(A_1^2 + A_2^2 + A_3^2 + A_4^2)^{1/2}$ does not change under a Lorentz transformation, i.e., the length of a 4-vector is an invariant quantity. Since A_4^2 is always negative, the length of a 4-vector can be zero even though it may have non-zero components.

A 4-vector is said to be **space-like, time-like** or **light-like** (also called **null vector**) according as the square of its length is positive, negative or zero. We shall use the subscripts 1, 2, 3, 4 to indicate the four components of a 4-vector while the components of a 3-vector would be specified by using Latin suffices x, y, z as subscripts. It should also be understood that the Greek suffices μ, ν, • • • will run from 1 to 4 while the Latin suffices i, j, • • • will run from 1 to 3.

It is customary to write a 4-vector A_μ as $A_\mu = (A_1, A_2, A_3, A_4)$ so that the square of its magnitude is given by

$$A_\mu A_\mu = A_1^2 + A_2^2 + A_3^2 + A_4^2\,.$$

Since, to a given observer, A_i's behave as the component of a 3-vector, we may write

$$A_\mu = (\mathbf{A}, A_4)$$

and $$A_\mu A_\mu = \mathbf{A}^2 + A_4^2 .$$

In particular, the position 4-vector x_μ is written as

$$x_\mu = (x_1, x_2, x_3, x_4) = (x, y, z, x_4) = (\mathbf{r}, x_4) = (\mathbf{r}, ict)$$

and its magnitude is given by the relation

$$x_\mu x_\mu = r^2 - c^2 t^2.$$

Interval and Light Cone

Let us now examine the relationship which various events may bear to each other. Consider two events occurring at points (x_1, y_1, z_1) and (x_2, y_2, z_2) at times t_1 and t_2 respectively in a frame of reference S. In the 4-dimensional space-time continuum these two events will be represented by the points (x_1, y_1, z_1, ict_1) and (x_2, y_2, z_2, ict_2) respectively. The distance Δs between these two points in 4-dimensional space is called the **interval** between the two events, and is given by

$$(\Delta s)^2 = (x_2 - x_1)^2 + (y_2 - y_1)^2 + (z_2 - z_1)^2 - c^2 (t_2 - t_1)^2 .$$

If we denote $(x_2 - x_1)^2$, etc. by Δx^2 ($\equiv (\Delta x)^2$) etc., this equation takes the form

$$(\Delta s)^2 = \Delta x^2 + \Delta y^2 + \Delta z^2 - c^2 \Delta t^2 = \Delta r^2 - c^2 \Delta t^2, \qquad (2.32)$$

where $\Delta r = + (\Delta x^2 + \Delta y^2 + \Delta z^2)^{1/2}$ is the spatial distance and Δt is the time interval between the two events. Equation (2.32) shows that, depending upon the spatial and temporal separations, the intervals between two events can be classified into the following three categories:

(1) $(\Delta s)^2 > 0$. Then $\Delta r^2 > c^2 \Delta t^2$, i.e., the spatial distance Δr between the two events is greater than the distance $c\,\Delta t$ which a ray of light can cover during the time interval Δt between the events. In this case, the two events are

either so far apart in coordinate space and/or occur in such rapid succession that one of them occurs even before a light signal from the other can reach it. Since no signal can be propagated with a velocity greater than that of light, the two events cannot be connected causally, i.e., one cannot be the cause of the other. The interval Δs between two such events is said to be a **space-like interval.**

(2) $(\Delta s)^2 < 0$. Then $\Delta r^2 < c^2 \Delta t^2$. In this case, the spatial distance between the two events can be covered by a light signal during the time interval between them. Therefore, one of the events can be the cause of the other, i.e., the two events can be connected causally. The interval Δs between two such events is said to be a **time-like interval.**

(3) $(\Delta s)^2 = 0$. Then $\Delta r^2 = c^2 \Delta t^2$ so that the spatial distance Δr between the two events can be just covered by a light signal during the time Δt and a causal relationship between the two events can occur. The interval Δs between two such events is said to be a **null** or **light-like interval.** It may be remarked that the path of a ray of light in space-time continuum is always represented by a null vector.

Since the length of a 4-vector in the world space is an invariant quantity, the interval Δs between any two given events is also invariant, i.e., Δs is independent of the frame of reference. Thus although both space and time intervals change under a Lorentz transformation, the space-time interval is still invariant. That is, the character of an interval between two events will not change under a Lorentz transformation. For instance, no Lorentz transformation can transform a space-like interval into a time-like interval, or vice versa. It may be noted that the most significant difference between the metrics of space and space-time, viz.,

$$(\Delta r)^2 = \Delta x^2 + \Delta y^2 + \Delta z^2$$

$$(\Delta s)^2 = \Delta x^2 + \Delta y^2 + \Delta z^2 - c^2 \Delta t^2$$

is that of *signature*. The space metric, irrespective of the choice of the coordinate system, is *positive definite* i.e., all the metric coefficients have positive signs which we represent by (+ + +). In contrast, the signature of the Minkowski space-time is (+ + + −) or, as we shall see later, (+ − − −). A signature of that kind is called *indefinite*. Consequently, whereas the interval between two distinct points in space is always positive, in space-time it can be positive, negative or even zero. Such a metric is sometimes called a **pseudo-metric.**

We shall now give a geometrical representation of the relation among the interval Δs between two events and the corresponding spatial and temporal intervals. According to equation (2.32), this relationship is given by

$$\Delta s^2 + c^2\,\Delta t^2 = \Delta r^2.$$

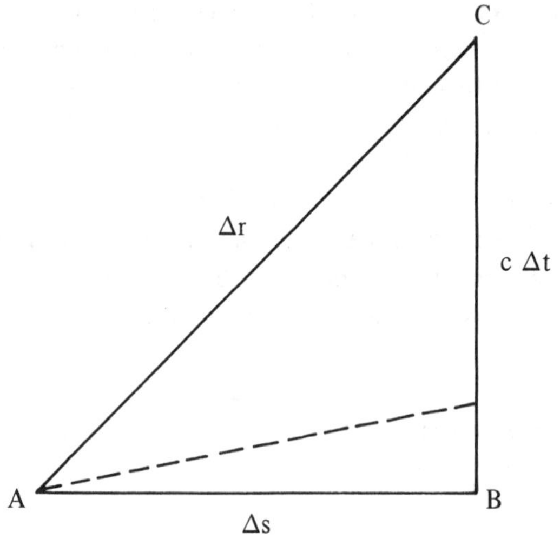

Fig. 2.6. Geometrical representation of Δs

The terms Δr^2 and $c^2\,\Delta t^2$ are always positive, so that, if the interval Δs is space-like, all the terms in this relation would be positive. Then the form of this relation suggests that Δr may be regarded as the hypotenuse of a right-angled triangle with Δs and $c\,\Delta t$ as its sides. This is shown in Fig. 2.6. Since Δs is invariant under a Lorentz transformation, the base of the triangle ABC would remain fixed in all inertial frames while its height and hypotenuse vary. The triangle ABC shows that for a fixed Δs, the quantities Δr and $c\,\Delta t$ can either increase together or decrease together; it is not possible to increase one of them and decrease the other, or to change only one of them and keep the other constant.

A similar geometrical interpretation can be given for a time-like interval.

For a null interval, the right-angled triangle collapses into a single vertical straight line equal to both Δr and $c\,\Delta t$.

We shall now illustrate the meaning of these terms by an example. Suppose that two events E_1 and E_2 occur in a laboratory at two points which are 9 m apart, the event E_2 occurring 10^{-8} s later than the event E_1. The spatial distance Δr between these two events is 9 m. The time interval Δt between these two events is 10^{-8} s and during this period light can travel a distance $c\,\Delta t = 3 \times 10^8 \times 10^{-8}$ m $= 3$ m only. This means that even a light signal emitted by the event E_1 at the time of its occurrence would reach the position of E_2 after the latter event has occurred. Since no signal can be propagated with a velocity greater than that of light and the event E_2 occurs before a light signal from E_1 reaches E_2, the two events E_1 and E_2 cannot be connected causally, i.e., E_1 cannot be the cause of E_2. Moreover, as $\Delta r = 9$ m is greater than $c\,\Delta t = 3$ m, the interval between the events E_1 and E_2 is space-like.

If the two events occur with a time interval of 10^{-6} s, then $\Delta r = 9$ m would be less than $c\,\Delta t = 3 \times 10^8 \times 10^{-6}$ m $= 300$ m. This interval would, therefore, be time-like. Since, in this case, signals from E_1 can reach E_2 before the event E_2 occurs, the two events can be connected causally.

If the time interval between the two events is 3×10^{-8} s, the interval Δs is null and the two events are just connected by a light signal.

Problem

What is the type of the interval between the emission and reception of a light pulse?

Let us again examine two events which occur at different times and at different places in a frame of reference S. The interval Δs between these two events is given by

$$\Delta s^2 = \Delta r^2 - c^2\,\Delta t^2. \tag{2.32'}$$

An interesting question arises: Does there exist a frame of reference in which these two events occur simultaneously? If such a frame of reference, say S′, does exist, then we must have

$$\Delta s'^2 = \Delta r'^2 - c^2\,\Delta t'^2 = \Delta r'^2,$$

because $\Delta t' = 0$ as the two events occur at the same time in S'. But $\Delta r'^2 > 0$ because $\Delta r'$ is real. Therefore the above equation gives

$$\Delta s'^2 > 0.$$

Since the interval is an invariant quantity, we conclude that

$$\Delta s^2 > 0.$$

This result shows that it is possible to find a frame of reference in which two given events occur at the same time provided that $\Delta s^2 > 0$, i.e., if the interval between the two given events is space-like. For a time-like or a null interval such a frame would not exist.

Another interesting question is: Does there exist a frame of reference in which these two events occur in the reverse order? In other words, is it possible to find a frame of reference, say S'', in which an event A occurring *earlier* than another event B in a frame S, occurs *later* than B?

If this were possible, the time interval $\Delta t''$ between these two events as observed in the frame S'' would be negative:

$$\Delta t'' < 0.$$

But $\Delta t''$ and Δt are connected by a Lorentz transformation:

$$\Delta t'' = \gamma \left(\Delta t - \frac{v}{c^2}\Delta x\right).$$

By virtue of the above equations,

$$\gamma \left(\Delta t - \frac{v}{c^2}\Delta x\right) < 0$$

or $\quad \Delta t < \frac{v}{c^2}\Delta x$, because γ is never negative.

or $$\Delta x^2 > \frac{c^4}{v^2}\Delta t^2 = \frac{c^2}{v^2} c^2 \Delta t^2$$

or $\quad \Delta x^2 > c^2 \Delta t^2$, because $c^2/v^2 > 1$

or $$\Delta x^2 - c^2 \Delta t^2 > 0.$$

But $\quad y^2 \geq 0, \ z^2 \geq 0.$

Combining the last three relations, we get

$$\Delta x^2 + \Delta y^2 + \Delta z^2 - c^2 \Delta t^2 > 0$$

or

$$\Delta s^2 > 0.$$

This shows that it is possible to find frames of reference in which two given events occur in the reverse order provided that the interval between the two events is space-like. For a time-like or null interval such frames would not exist.

To put it all together, to observers in different inertial frames, two events A and B connected by a space-like interval can occur simultaneously, or with A occurring earlier than B, or A occurring later than B. Thus for space-like intervals, the words *simultaneous*, *earlier* and *later* have no sense.

On the other hand, as we have just seen, for a time-like interval if two events A and B occur in a frame of reference at different times in a certain order, then it is impossible to find a frame of reference in which these two events occur either simultaneously or in the reverse order. Consequently the words *earlier*, *simultaneous* and *later* have absolute significance for events separated by time-like intervals. For such intervals, the sequence of events in time is preserved for observers in different inertial frames. The same is true for null intervals.

Now consider the motion of a material particle. In the space-time continuum, it will describe a world line, each point on the line representing an event. Since the velocity of a material particle is always less than the velocity of light, any two points on its world line can always be connected causally. Hence the interval between any two events on the world line of a material particle must be time-like. For such an interval the order of the events has an absolute significance. Therefore, for instance, it is not possible to find a frame of reference in which a man is observed to reap his harvest before he sows it. Similarly, it is impossible to find an observer for whom a bullet strikes the target before it leaves the gun.

Problem

Show that for the two events occurring at different times and places, there exists a frame of reference in which these two events occur at the same place, provided the interval between the events is time-like.

Let us now represent these conclusions graphically. The set of events which are time-like or light-like with respect to any event O is said to constitute the light cone of the event O. To visualize this cone, consider a 2-dimensional space-time diagram like Fig. 2.7. Such diagrams are sometimes called **Minkowski's diagrams.** The horizontal line is taken as the x-axis while the vertical line is the ct-axis. The time axis is represented by ct in preference to t merely for the sake of dimensional homogeneity, as the product of velocity and time is length. The given event O is taken as the origin of this coordinate system. In particular, we shall assume that O is a light pulse which in 3-dimensional space immediately expands in all directions with a velocity c. Then in the space-time diagram, it would make an angle θ with the ct-axis such that $\tan\theta = \frac{x}{ct} = 1$. This yields $\theta = 45°$. The line AOB in the figure represents the world line of such a particle. The same is true for the line COD. Thus any event on these lines is a light-like event with respect to the event O. Since a material particle would always be moving with a velocity $v<c$, its world line will make an angle of less than 45° with the ct-axis and therefore its motion at all times will be represented by a line lying in the sectors AOC

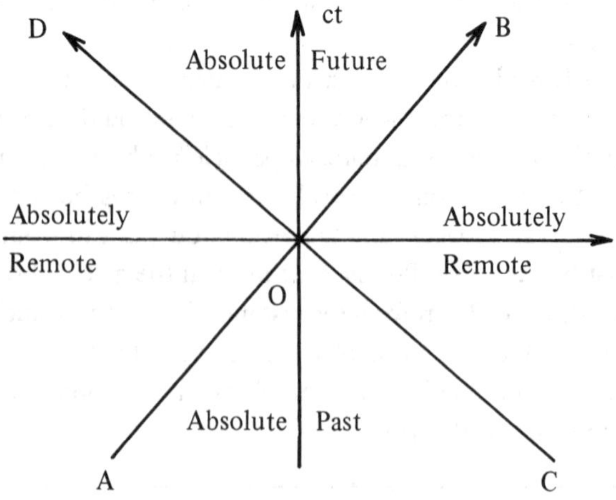

Fig. 2.7. Past, future and absolutely remote

and DOB. All the events in the sector AOC will be in the past of the event O. Similarly, all the events in the sector DOB will be in the future of O. Since for time-like and light-like intervals the words *before* and *after* have absolute significance, all the events in the sector BOD are future events relative to the event O in all the reference systems while all the events in the sector AOC are past events relative to the event O in all the reference
systems. The sectors AOC and BOD therefore represent events which lie respectively in the *Absolute Past* and *Absolute Future* of the event O. Events in the sector BOD can be influenced by what happens at O, while signals from the events in the sector AOC could reach the origin at or before $t = 0$ and thus influence what happens at O.

Next consider the sectors DOA and BOC. Events in these sectors cannot be connected causally to the event O because no signal can proceed with a velocity greater than that of light. The intervals are therefore space-like. The events in the two sectors are said to be *absolutely remote* from the event O. Any event in these regions can be made simultaneous with, before or later than O depending upon the frame of reference used.

If we consider all the three space coordinates x, y, z then the three regions of *past*, *future* and *absolutely remote* are separated from one another by the hypercone

$$x^2 + y^2 + z^2 - c^2 t^2 = 0 \tag{2.33}$$

in the 4-dimensional space. The axis of the hypercone described by equation (2.33) coincides with the ct-axis. This hypercone is called the **null cone** or the **light cone** because a light signal transmitted from O will have its world line on its surface. A light cone is shown in Fig. 2.8.

If the two events E_1 and E_2 are separated by time-like intervals from O, it is not necessary that the interval between them should also be time-like. In fact the interval between E_1 and E_2 will be time-like only if the world line connecting them makes an angle of less than 45° with the time axis.

Problem

Write down equations of the light-lines AOB and COD of Fig. 2.7 and show that they are perpendicular to each other. Will these lines be common for all observers?

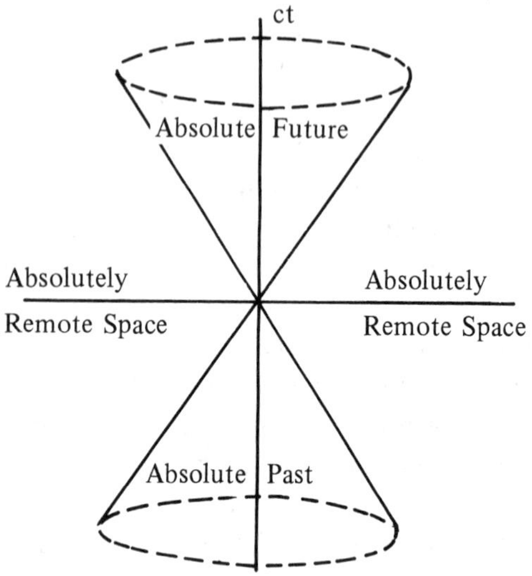

Fig. 2.8. Light cone

Proper Time

Consider a particle moving with a velocity **V** in an inertial frame of reference. Let (x, y, z) and (x + dx, y + dy, z + dz) represent two positions of the particle at times t and t + dt respectively; the times being noted on a clock fixed in that frame. These two positions of the particle at times t and t + dt may be considered as two events represented by the points (x, y, z, ict) and (x + dx, y + dy, z + dz, ic (t + dt)) in a 4-dimensional space-time continuum; each event may be characterized by the emission of a spark. Then the interval ds between these two events will be given by

$$ds^2 = dx^2 + dy^2 + dz^2 - c^2\, dt^2. \qquad (2.34)$$

Since ds^2 is an invariant quantity, in the frame moving with the particle, it would be given by

$$ds^2 = -\, c^2\, d\tau^2, \qquad (2{:}35)$$

having the same space coordinates. Here $d\tau$ is the time interval noted on a clock fixed in a frame moving with the particle. From equations (2.34) and (2.35), we have

$$-c^2 d\tau^2 = dr^2 - c^2 dt^2$$

or
$$d\tau^2 = dt^2 - \frac{1}{c^2} dr^2$$
$$= dt^2 [1 - \frac{1}{c^2}(\frac{dr}{dt})^2]$$

or
$$d\tau = dt \sqrt{1 - V^2/c^2} . \quad (2.36)$$

The time interval $d\tau$ is called the **proper time interval**. Integrating equation (2.36), we get an expression for the proper time τ:

$$\tau = \int dt \sqrt{1 - V^2/c^2} . \quad (2.37)$$

Equation (2.37) shows that for $V \ll c$, the proper time reduces to ordinary time.

It may be stressed that relations (2.36) and (2.37) are true for the accelerated as well as unaccelerated frames.

Tensor Formulation of a Physical Law

According to the special theory of relativity, equations expressing a law of nature must have the same form in all inertial frames and therefore should be covariant under a Lorentz transformation. Since a Lorentz transformation is an orthogonal transformation in 4-dimensional space-time continuum, a 4-tensor equation, as it retains its form under an orthogonal transformation (see Appendix C), will not change its form under a Lorentz transformation. Hence it should be possible to write equations expressing a law of nature in 4-tensor form.

The Poincaré Group

A linear transformation in space-time continuum, viz.,

$$x'_\mu = a_{\mu\nu} x_\nu + b_\mu, \quad \mu, \nu = 1, 2, 3, 4 \quad (2.38)$$

where the space coordinates $x_1 \equiv x$, $x_2 \equiv y$, $x_3 \equiv z$ are real, and $x_4 = ict$ is pure imaginary, while $a_{\mu\nu}$ and b_μ are constants, is said to be a Lorentz transformation if it satisfies the equation

$$dx_1^2 + dx_2^2 + dx_3^2 - c^2\, dt^2 = dx_1'^2 + dx_2'^2 + dx_3'^2 - c^2\, dt'^2. \qquad (2.39)$$

Equation (2.39) may be written as

$$dx_1^2 + dx_2^2 + dx_3^2 + dx_4^2 = dx_1'^2 + dx_2'^2 + dx_3'^2 + dx_4'^2$$

or

$$dx_\mu\, dx_\mu = dx'_\mu\, dx'_\mu. \qquad (2.40)$$

Taking differentials of both sides of equation (2.38), we get

$$dx'_\mu = a_{\mu\nu}\, dx_\nu.$$

Substituting this expression for dx'_μ in equation (2.40), we have

$$dx_\mu\, dx_\mu = a_{\mu\nu}\, dx_\nu\, a_{\mu\sigma}\, dx_\sigma = a_{\mu\nu}\, a_{\mu\sigma}\, dx_\nu\, dx_\sigma, \qquad (2.41)$$

where, in order to avoid any confusion which may arise if any letter index occurs more than twice, we have used different dummy indices on the right hand side of equation (2.41). Comparing the coefficients on the two sides of equation (2.41), we get

$$a_{\mu\nu}\, a_{\mu\sigma} = \delta_{\nu\sigma}. \qquad (2.42)$$

Since x_1, x_2, x_3 are all real and x_4 is pure imaginary, the coefficients $a_{\mu\nu}$ and the constants b_μ should be such that a_{ij}, a_{44}, b_i are real while a_{i4}, a_{4i}, b_4 are pure imaginary. Equation (2.42) expresses the orthogonality of the transformation defined by equation (2.38).

Thus the linear transformation $x'_\mu = a_{\mu\nu}\, x_\nu + b_\mu$ is a Lorentz transformation provided that $a_{\mu\nu}\, a_{\mu\sigma} = \delta_{\nu\sigma}$ and a_{ij}, a_{44}, b_i are real while a_{i4}, a_{4i}, b_4 are pure imaginary.

Equations (2.38) and (2.39) may be written in matrix form as

$$X' = A\, X + b \qquad (2.43)$$

$$dX^T\,dX = dX'^T\,dX', \tag{2.44}$$

where

$$X = \begin{pmatrix} x_1 \\ x_2 \\ x_3 \\ x_4 \end{pmatrix}, \qquad X' = \begin{pmatrix} x'_1 \\ x'_2 \\ x'_3 \\ x'_4 \end{pmatrix}, \qquad b = \begin{pmatrix} b_1 \\ b_2 \\ b_3 \\ b_4 \end{pmatrix},$$

$$A = \begin{pmatrix} a_{11} & a_{12} & a_{13} & a_{14} \\ a_{21} & a_{22} & a_{23} & a_{24} \\ a_{31} & a_{32} & a_{33} & a_{34} \\ a_{41} & a_{42} & a_{43} & a_{44} \end{pmatrix}$$

and X^T stands for the transpose of the matrix X. The matrix A is called the **transformation matrix.**

Differentiating equation (2.43), we get

$$dX' = A\,dX. \tag{2.45}$$

Taking the transpose of both sides of equation (2.43), we obtain

$$X'^T = X^T A^T + b^T.$$

Taking the differentials of both sides of this equation, we have

$$dX'^T = dX^T A^T. \tag{2.46}$$

Substituting the expressions for dX'^T and dX' from equations (2.45) and (2.46) in equation (2.44), we obtain $dX^T\,dX = dX^T A^T A\,dX$, so that $A^T A = I$, i.e., the transformation matrix A is orthogonal. Thus, in matrix notation, a Lorentz transformation may be defined as a linear transformation

$$X' = A\,X + b$$

in the Minkowski-space where the transformation matrix A is orthogonal, $A^T A = I$, and the elements of A and b are such that a_{ij}, a_{44}, b_i are real while a_{i4}, a_{4i}, b_4 are pure imaginary.

It may be pointed out that the Lorentz transformations are a subset of orthogonal transformations in a 4-dimensional space, where x_1, x_2, x_3, x_4 are, in general, complex variables, so that the conditions that

a_{ij}, a_{44}, b_1 are real

and a_{i4}, a_{4i}, b_4 are pure imaginary

are not imposed. Symbolically we can write

$$A = \begin{pmatrix} r & r & r & i \\ r & r & r & i \\ r & r & r & i \\ i & i & i & r \end{pmatrix}, \qquad b = \begin{pmatrix} r \\ r \\ r \\ i \end{pmatrix},$$

where r stands for a real and i for a pure imaginary element.

By taking the determinant of both sides of the relation $A^T A = I$, it follows that $|A^T A| = |I|$ or $|A^T|\ |A| = |A|\ |A| = |A|^2 = 1$, where we have made use of the fact that the determinant $|A|$ of a matrix A is unchanged by the interchange of its rows and columns. This yields $|A| = \pm 1$. A Lorentz transformation with $|A| = +1\ (-1)$ is called a **proper (improper)** Lorentz transformation.

It is also common to use the variable $x_0 = ct$. In this notation, a Lorentz transformation is defined as a linear transformation

$$x'_\mu = a_{\mu\nu}\, x_\nu + b_\mu, \qquad \mu, \nu = 0, 1, 2, 3$$

satisfying the equation

$$dx_1^2 + dx_2^2 + dx_3^2 - c^2\, dt^2 = dx_1'^2 + dx_2'^2 + dx_3'^2 - c^2\, dt'^2 . \qquad (2.39')$$

Equation (2.39) may then be written as

$$- dx_0^2 + dx_1^2 + dx_2^2 + dx_3^2 = - dx_0'^2 + dx_1'^2 + dx_2'^2 + dx_3'^2. \qquad (2.47)$$

To express the definition in terms of matrices, we define the column vectors X′ and X and the matrices g and A by the equations

$$X = \begin{pmatrix} x_0 \\ x_1 \\ x_2 \\ x_3 \end{pmatrix}, \quad X' = \begin{pmatrix} x'_0 \\ x'_1 \\ x'_2 \\ x'_3 \end{pmatrix},$$

$$g = \begin{pmatrix} 1 & 0 & 0 & 0 \\ 0 & -1 & 0 & 0 \\ 0 & 0 & -1 & 0 \\ 0 & 0 & 0 & -1 \end{pmatrix} = \text{diag}\,(1, -1, -1, -1),$$

$$A = [a_{\mu\nu}] = \begin{pmatrix} a_{00} & a_{01} & a_{02} & a_{03} \\ a_{10} & a_{11} & a_{12} & a_{13} \\ a_{20} & a_{21} & a_{22} & a_{23} \\ a_{30} & a_{31} & a_{32} & a_{33} \end{pmatrix}.$$

Then, in matrix form, equations (2.38) and (2.47) may be written as

$$X' = A\,X + b \tag{2.43'}$$

$$dX^T g\, dX = dX'^T g\, dX'. \tag{2.48}$$

Substituting the expressions for dX'^T and dX' obtained by using equation (2.43) in equation (2.48), we get

$$dX^T g\, dX = dX^T A^T g\, A\, dX.$$

Since this relation must be valid for all dX, it would be possible only if

$$A^T g\, A = g.$$

Notice that now all the a_μ and the b_μ are real. Hence a Lorentz transformation may be defined as a linear transformation, $X' = A X + b$, in a space-time continuum $(x_0 = ct, x_1, x_2, x_3)$ such that $A^T g A = g$, where $g = \text{diag}(1, -1, -1, -1)$ and all the elements of A and b are real.

We will now show that the Lorentz transformations form a group under successive application of transformations, i.e., they possess the following four characteristics of a group.

(1) The product of any two Lorentz transformations

$$X' = A X + b, \quad A^T g A = g \tag{2.49}$$

$$X'' = C X' + d, \quad C^T g C = g \tag{2.50}$$

is again a Lorentz transformation. That is if the Lorentz transformation (2.49) takes us from reference system S to S′, and the Lorentz transformation (2.50) takes us from the reference system S′ to S″, then the successive applications of these transformations will give another Lorentz transformation which will take us from the reference system S to S″. This resultant Lorentz transformation is obtained by substituting the expression for X' from the first relation in the second:

$$X'' = C(A X + b) + d = C A X + C b + d = F X + h, \tag{2.51}$$

where $F = C A$ and $h = C b + d$.

Since $F^T g F = (C A)^T g (C A) = A^T C^T g C A = A^T g A = g$, where the relations $A^T g A = g = C^T g C$ have been used, we conclude that the transformation (2.51) is a Lorentz transformation.

It may be noticed that these transformations are not commutative because in general $A C \neq C A$, i.e., we arrive at different results depending upon whether the transformation represented by A is applied after that represented by C, or vice versa.

(2) The multiplication is associative because of the linearity of the transformations.

(3) The Lorentz transformation

$$X' = X = I X, \qquad I^T g I = g$$

serves as the identity transformation, i.e., as a transformation which leaves everything unchanged.

(4) The inverse of each Lorentz transformation exists and is itself a Lorentz transformation. Consider the Lorentz transformation

$$X' = A\,X + b, \qquad A^T g\,A = g. \tag{2.49'}$$

Since $|A| = \pm 1 \neq 0$, the matrix A is non-singular and its inverse exists. Multiplying both sides of equation (2.49′) by A^{-1} from the left, we get

$$A^{-1} X' = X + A^{-1} b$$

or

$$X = A^{-1} X' - A^{-1} b = A^{-1} X' + d, \quad \text{where } d = A^{-1} b. \tag{2.52}$$

To find out whether this linear transformation is a Lorentz transformation or not, we consider the matrix relation

$$A^T g\,A = g.$$

Multiplying from the left by $(A^T)^{-1}$ and from the right by A^{-1}, we get

$$(A^T)^{-1} A^T g\,A\,A^{-1} = (A^T)^{-1} g\,A^{-1}$$

or

$$g = (A^{-1})^T g\,A^{-1}.$$

This shows that the linear transformation (2.52) is a Lorentz transformation. This is the inverse of the Lorentz transformation (2.49′) and it takes us from S′ back to S.

Hence the set of all Lorentz transformations forms a group. This group is called the **Poincaré group** or the **inhomogeneous Lorentz group** or the **general Lorentz group** or the **complete Lorentz group** and is denoted by $\mathcal{L}$.

Special Cases of the Lorentz Transformations

We will now consider some special cases of the Lorentz transformations.

1. Space-Time Rotations

For b = 0, equations (2.49) reduce to $X' = AX$, $A^T g A = g$. This transformation represents rotations in space-time continuum. The set of such transformations form a subgroup of $\mathscr{L}$. This subgroup is called the **homogeneous Lorentz group** or the **Lorentz group** and is denoted by L. In this case any matrix A satisfying the equation $A^T g A = g$ is said to define a Lorentz transformation. A Lorentz transformation with A as transformation matrix will frequently be mentioned as the Lorentz transformation A.

The **restricted Lorentz transformations**

$$x' = \gamma (x - v t)$$

$$y' = y$$

$$z' = z$$

$$t' = \gamma (t - \frac{v}{c^2} x)$$

form a subgroup of the Lorentz group L under successive application of transformations. Using $x_0 = ct$, these transformations may be written as

$$x_0' = \gamma (x_0 - \frac{v}{c} x_1)$$

$$x_1' = \gamma (x_1 - \frac{v}{c} x_0)$$

$$x_2' = x_2$$

$$x_3' = x_3.$$

Notice the symmetry in transformation equations for x_0' and x_1'. The transformation matrix A(v) is then given by

$$A(v) = \begin{pmatrix} \gamma & -\gamma \frac{v}{c} & 0 & 0 \\ -\gamma \frac{v}{c} & \gamma & 0 & 0 \\ 0 & 0 & 1 & 0 \\ 0 & 0 & 0 & 1 \end{pmatrix}$$

Since the matrix A is symmetric, i.e., $A^T = A$, we have $A^T g A = A g A = g$ by actual calculations, showing that these transformations also form a group. The det A is equal to +1 showing that it is a proper Lorentz transformation.

Problem

Show that the Lorentz transformations without rotation do not form a subgroup of L because the product of two such transformations is a transformation *with rotation*.

---

2. *Translations in Space-Time Continuum*

For A = I, equations (2.49) reduce to

$$X' = I X + b = X + b, \quad I^T g I = g$$

or

$$x'_\mu = x_\mu + b_\mu.$$

These transformations represent linear displacement of the origin in space-time continuum. The determinant of the transformation matrix is again +1 showing that it is also a proper Lorentz transformation. It can be easily checked that these transformations also form a group under successive application of transformations. This is a subgroup of the general Lorentz group $\mathcal{L}$ and is denoted by T.

3. *Space-inversion*

The transformation

$$x' = -x, \quad y' = -y, \quad z' = -z, \quad t' = t,$$

which changes the sign of every space coordinate, also satisfies equation (2.39) and is called **space-inversion**. This Lorentz transformation may be put in matrix form as

$$X' = A X, \quad \text{where} \quad A = \text{diag}\,(1, -1, -1, -1) = g$$

so that

$$A^T g A = g^T g g = g g g = g^3 = g.$$

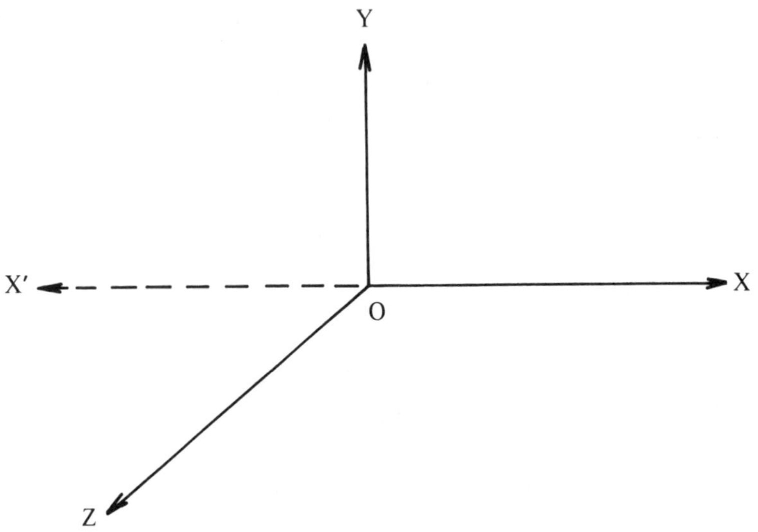

Fig. 2.9. Space-reflection of x-axis

Since $|A| = -1$, this is an improper Lorentz transformation.

To understand its physical meaning, consider a right-handed coordinate system OX, OY, OZ and imagine the reflection of x-axis in the yz plane (Fig. 2.9) which may be regarded as a mirror placed at right angles to the x-axis. This reflection of x-axis changes x to $-x$ and for this reason is called the mirror reflection of x. This transformation, $x' \to -x$, $y' \to y$, $z' \to z$, $t' \to t$, is called space-reflection. Then OX$'$, OY, OZ form a left-handed coordinate system. That is a reflection changes a right-handed coordinate system to a left-handed one and vice versa. If we consider the reflection of each axis in a plane perpendicular to that axis, we obtain the space-inversion

$$x' = -x, \quad y' = -y, \quad z' = -z, \quad t' = t,$$

which also changes a right-handed coordinate system to a left-handed coordinate system and vice versa. However, two left-handed coordinate systems, one obtained by space-reflection and the other by space-inversion of the same right-handed coordinate system, are not the same.

If we consider the reflection of two space axes, say $y \to -y$ and $z \to -z$, then the final effect is equivalent to a rotation of the yz plane about the x-axis through 180° (Fig. 2.10). Therefore such a reflection of space axes is no inversion at all. Hence, space-inversion is equivalent to space-reflection of a single axis, say x-axis, plus a rotation of the yz-plane about x-axis through 180°.

We have seen that this Lorentz transformation, viz., space-inversion, relates the space and time coordinates of a point in a right-handed and a left-handed coordinate system. Since the equations expressing a law of physics

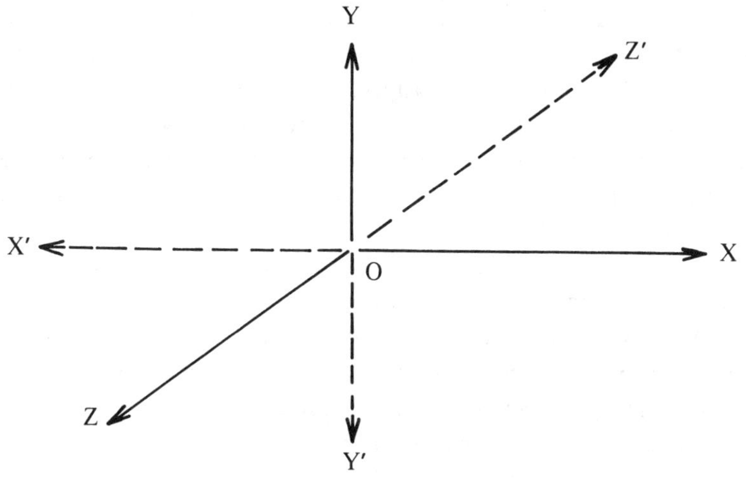

Fig. 2.10. Space-inversion

retain their form under a Lorentz transformation, laws of physics should remain invariant under space-inversion so that it should not be possible to distinguish between a right-handed and a left-handed coordinate system by any physical experiment. Since the left-handed coordinate system obtained from a right-handed coordinate system by space-inversion is its mirror image, the mirror image of a physical situation should be realizable in nature with equal probability. On the other hand, if space-inversion is completely violated, the mirror image of a physical phenomenon would not occur in nature. If the space-inversion is violated to a certain degree only, a phenomenon and its mirror image will occur in nature with different probabilities, the ratio of their probabilities depending upon the degree of violation of the space-inversion.

On the suggestion of two physicists, Lee and Yang, an experiment was performed by Wu *et al.*[2] in 1957, which showed that for weak interactions the space-inversion is not a valid operation.

4. Time-reversal

The linear transformation

$$x' = x, \quad y' = y, \quad z' = z, \quad t' = -t$$

also satisfies equation (2.39) and is therefore a Lorentz transformation. It is called time-reversal and interchanges past and future. The determinant of the transformation matrix is −1; it is therefore an improper Lorentz transformation. If time-reversal transformation is a valid operation, say, for the motion of an object, it will be equally permissible to carry out the motion of this object in the reverse order, like a movie played backward. If time-reversal is violated, such a motion of the object should not be possible. The validity of the time-reversal is in doubt at present.

It seems that the laws of physics are essentially invariant only under proper Lorentz transformations as originally envisaged by Einstein.

We may remark that an erroneous impression has been created that space-inversion turns space inside out while time-reversal makes time run backwards. Such interpretations stem from a wish to understand these transformations in purely geometrical terms.

Subsets of Lorentz Transformations

The Lorentz transformations may be divided into four subsets by studying the possible values for the determinant of A, viz., $|A|$ and the element $a_{00} \equiv A_{00}$ of the matrix A defining the Lorentz transformation.

For that purpose, we shall first show that even with the variables x_0, x_1, x_2, x_3, the determinant of the transformation matrix A of a Lorentz transformation can be either + 1 or − 1. For such a transformation, the matrix A must satisfy the equation

$$A^T g A = g.$$

By taking the determinants of both sides, we get

$$|A^T| \; |g| \; |A| = |g|.$$

Since $|g| = -1 \neq 0$, it can be cancelled from both sides, so that

$$|A^T| \; |A| = 1 \qquad \text{or} \qquad |A| = \pm 1.$$

This is the required result.

We shall next write the 00 element of $A^T g A = g$. From this equation, we have

$$(A^T g A)_{00} = g_{00}$$

or $$A^T_{0\mu} g_{\mu\nu} A_{\nu 0} = g_{00}, \qquad \text{where } \mu, \nu = 0, 1, 2, 3 .$$

Summing over ν, we get

$$A^T_{0\mu} (g_{\mu 0} A_{00} + g_{\mu 1} A_{10} + g_{\mu 2} A_{20} + g_{\mu 3} A_{30}) = g_{00}.$$

Summing over μ, and noting that g is a diagonal matrix, we obtain

$$A^T_{00} g_{00} A_{00} + A^T_{01} g_{11} A_{10} + A^T_{02} g_{22} A_{20} + A^T_{03} g_{33} A_{30} = g_{00}.$$

Substituting the expressions for $g_{\mu j}$, the elements of the diagonal matrix g, we get

$$- A^2_{00} + A^2_{10} + A^2_{20} + A^2_{30} = -1$$

or $$A^2_{00} = 1 + A^2_{10} + A^2_{20} + A^2_{30}.$$

Since all the elements of A are real, this gives $A^2_{00} \geq 1$, i.e., either $A_{00} \geq 1$ or $A_{00} \leq -1$.

The conditions

1. $|A| = \pm 1$ and
2. either $A_{00} \geq 1$ or $A_{00} \leq -1$

determine the four pieces of *homogeneous Lorentz transformations*. These are tabulated below:

Piece	1	2	3	4
$\lvert A \rvert$	+ 1	− 1	− 1	+ 1
$a_{00} \equiv A_{00}$	≥ 1	≥ 1	≤ -1	≤ -1
	$L^{\uparrow}_{+}$	$L^{\uparrow}_{-}$	$L^{\downarrow}_{-}$	$L^{\downarrow}_{+}$

These four pieces are respectively denoted by $L_+^\uparrow$, $L_-^\uparrow$, $L_-^\downarrow$, $L_+^\downarrow$ and are shown in the same table. The + or − subscript in each L tells whether the determinant of the transformation matrix is + 1 or − 1. The arrow in each L points upwards or downwards according as $A_{00} \geq 1$ or ≤ -1.

Let us next find out to which pieces the space-inversion and the time-reversal operations (transformations) belong.

The space-inversion or parity operation is defined by

$$t' = t, \;\; x' = -x, \;\; y' = -y, \;\; z' = -z,$$

so that the parity operator P is given by

$$P = \text{diag}\,(1, -1, -1, -1) = -g.$$

The determinant of P is equal to minus one, $|P| = -1$, and $P_{00} = +1$. This shows that the parity operator P belongs to $L_-^\uparrow$.

The time-reversal operation is defined by

$$t' = -t, \;\; x' = x, \;\; y' = y, \;\; z' = z$$

so that the time-reversal operator T is given by

$$T = \text{diag}\,(-1, 1, 1, 1) = g.$$

The determinant of T is minus one, $|T| = -1$, and $T_{00} = -1$. This means that T belongs to $L_-^\downarrow$. It may be noted that the operator PT is given by

$$PT = \text{diag}\,(-1, -1, -1, -1) = -I$$

so that $|PT| = 1$ and $(PT)_{00} = P_{0\nu}\, T_{\nu 0} = P_{00}\, T_{00} = -1$, because all other elements of $P_{0\nu}\, T_{\nu 0}$ vanish. This shows that PT belongs to $L_+^\downarrow$.

PROBLEMS

1. Calculate the speed with which a car must move in order that its length may be shortened to half of its proper length.
2. A car moves at a speed of 160 km per hour. If the length of the car is 2.4 m, calculate the decrease in its length as noted by a stationary observer.
3. Two events occur simultaneously at points (21, 2, 1) and (1, 0, 0) of a frame S. Determine the time-interval between them in a frame S' moving with a speed 0.6 c relative to S and along the positive direction of their common x-axis.
4. Two electrons are moving in opposite directions each with uniform speed of 0.7 c relative to the laboratory frame. What is the relative speed of the electrons as measured in the rest frame of one of the electrons?
5. Λ-hyperon has an average life-time of about 2.5×10^{-10} s in its rest frame. If it travels a distance of 0.05 m in a bubble chamber before decaying, calculate its velocity.

3

RELATIVISTIC MECHANICS

According to the special theory of relativity, equations expressing a physical law must have the same form in all inertial frames of reference, i.e., they should be covariant under a Lorentz transformation. Although Newton's laws of motion have been remarkably successful in describing the motion of objects moving with velocities which are very small as compared with the velocity of light, they are not invariant under a Lorentz transformation. Newtonian mechanics is, therefore, not consistent with the special theory of relativity. In order that mechanics may be consistent with the special relativity, we have to seek a new law of motion which must be such that the equations expressing this law are not only covariant under a Lorentz transformation but also reduce to Newton's equations of motion whenever the velocity V of the object is very small as compared to the velocity of light. The first requirement follows from the postulates of special relativity while the second one is a consequence of the fact that the results derived by using Newton's laws are consistent with experiment for all those phenomena in which $V \ll c$.

Problem

Show that Newton's equations of motion are not covariant under a Lorentz transformation.

To find the Lorentz covariant equations of motion of a particle under a given external force, we make use of the elegant 4-dimensional technique and introduce position, velocity, momentum and force through 4-vectors. Since a 4-vector equation is always covariant, this technique guarantees that the new equations of motion will have the same form in all inertial frames of reference. Moreover, we will introduce different 4-vectors such that for velocities which are very small as compared to that of light, the space components of each 4-vector reduce to the corresponding classical expressions. This will ensure that for $V \ll c$, new equations of motion reduce to the equations expressing Newton's laws of motion.

Position 4-Vector and 4-Velocity

We start our search for correct equations of motion by considering the motion of a particle moving with a velocity $\mathbf{V} = (V_x, V_y, V_z)$. The position (x_1, x_2, x_3) of this particle at any instant t can be described by a point $P(x_1, x_2, x_3, x_4 = ict)$ in a frame of reference K in 4-dimensional space-time continuum. The vector drawn from the origin O(0, 0, 0, 0) of K to the position of the particle is called its position 4-vector. In this case, the coordinates x_1, x_2, x_3, x_4 of the particle are also the components of this 4-vector. The position 4-vector is therefore usually denoted by x_μ. The rate of change of position 4-vector x_μ of the particle with respect to its proper time will also be a 4-vector because the proper time is an invariant quantity. This is called 4-velocity and is denoted by V_μ:

$$V_\mu = \frac{dx_\mu}{d\tau}. \tag{3.1}$$

Problem

Can we obtain a velocity 4-vector by differentiating x_μ with respect to t?

Equation (3.1) may be written as

$$V_\mu = \frac{dx_\mu}{dt}\frac{dt}{d\tau} = \frac{1}{\sqrt{1 - V^2/c^2}}\frac{dx_\mu}{dt} = \gamma(V)\frac{dx_\mu}{dt}.$$

Let us write explicitly the components of the 4-velocity V_μ. We have

$$V_1 = \gamma(V)\frac{dx_1}{dt} = \gamma(V)\, V_x. \tag{3.2a}$$

Similarly,

$$V_2 = \gamma(V)\, V_y \tag{3.2b}$$

$$V_3 = \gamma(V)\, V_z \tag{3.2c}$$

$$V_4 = \gamma(V)\frac{dx_4}{dt} = ic\,\gamma(V). \tag{3.2d}$$

We may therefore write

$$V_\mu = (V_1, V_2, V_3, V_4) = \gamma(V)\,(\mathbf{V}, ic) = \frac{1}{\sqrt{1 - V^2/c^2}}\,(\mathbf{V}, ic). \qquad (3.2e)$$

For $V/c \to 0$, the function $\gamma(V) \to 1$, and the first three components of the 4-velocity reduce to the components of the Newtonian 3-velocity. The magnitude of the 4-velocity is given by

$$V_\mu V_\mu = V_1^2 + V_2^2 + V_3^2 + V_4^2 = \gamma(V)\,(V_x^2 + V_y^2 + V_z^2) - c^2\gamma^2(V)$$

$$= \gamma^2(V)\,(V^2 - c^2) = \frac{1}{1 - V^2/c^2}(V^2 - c^2) = -\,c^2. \qquad (3.3)$$

This is a constant quantity as the magnitude of every 4-vector must be. This also shows that the 4-velocity is a time-like vector.

Since V_μ is a 4-vector, its components V_1, V_2, V_3, V_4 transform in the same way as x_1, x_2, x_3, x_4, i.e., as x, y, z, ict. Hence the transformation equations for velocity are given by

$$V_1' = \gamma(v)\,(V_1 + i\frac{v}{c}V_4) \qquad (3.4a)$$

$$V_2' = V_2 \qquad (3.4b)$$

$$V_3' = V_3 \qquad (3.4c)$$

$$V_4' = \gamma(v)\,(V_4 - i\frac{v}{c}V_1). \qquad (3.4d)$$

Here V and V' are the speeds of the particle in the two frames of reference, v is the relative speed of the two frames and $\gamma(v) = 1/\sqrt{1 - v^2/c^2}$. To find the transformation laws for V_x, V_y, V_z, we substitute the expressions for V_1, V_2, V_3 from equations (3.2) in equations (3.4) and use similar relations between V_1', V_2', V_3' and V_x', V_y', V_z'. We then obtain

$$V_x' = A\,(V_x - v) \qquad (3.5a)$$

$$V_y' = \gamma(V)\,V_y/\gamma(V') = A\,V_y/\gamma(v) \qquad (3.5b)$$

$$V'_z = \gamma(V)\,V_z/\gamma(V') = A\,V_z/\gamma(v), \qquad (3.5c)$$

where $A = \gamma(v)\,\gamma(V)/\gamma(V')$. $\qquad (3.5d)$

Now substituting the expressions for V_1, V_4 and V'_4 from equations (3.2a), (3.2d) and from a primed equation similar to equation (3.2d) in equation (3.4d), we get

$$1 = \gamma(v)\, \gamma(V)\, (1 - \frac{v}{c^2} V_x)/\gamma(V').$$

Using equation (3.5d), this can be written as

$$1 = A\,(1 - \frac{v}{c^2} V_x). \tag{3.6}$$

Substituting the expression for A from equation (3.6) in equations (3.5a, b, c), we get

$$V'_x = \frac{V_x - v}{1 + \frac{v}{c^2}\, V_x} \tag{3.7a}$$

$$V'_y = \frac{V_y}{\gamma\,(1 + \frac{v}{c^2}\, V_x)} \tag{3.7b}$$

$$V'_z = \frac{V_z}{\gamma\,(1 + \frac{v}{c^2}\, V_x)}, \tag{3.7c}$$

where, as already pointed out, γ stands for $\gamma(v)$.

The last three equations give the transformation law for the components of 3-velocity.

4-Momentum and Variation of Mass with Velocity

Let us consider a particle in motion. Let V_μ be the 4-velocity of the particle. Then the product of the velocity 4-vector with a 4-scalar m_0 will again be a 4-vector which we denote by p_μ:

$$p_\mu = m_0\, V_\mu\, .$$

This 4-vector is said to represent the 4-momentum of the particle. By virtue of equation (3.2e), we can write it as

$$p_\mu = m_0\, V_\mu = \frac{m_0}{\sqrt{1 - V^2/c^2}}\, (\mathbf{V}, ic) \tag{3.8}$$

so that

$$p_1 = m_0 V_1 = \frac{m_0}{\sqrt{1 - V^2/c^2}} V_x \tag{3.9a}$$

$$p_2 = m_0 V_2 = \frac{m_0}{\sqrt{1 - V^2/c^2}} V_y \tag{3.9b}$$

$$p_3 = m_0 V_3 = \frac{m_0}{\sqrt{1 - V^2/c^2}} V_z \tag{3.9c}$$

$$p_4 = m_0 V_4 = \frac{m_0}{\sqrt{1 - V^2/c^2}} ic\ . \tag{3.9d}$$

The first three of these equations show that if we want the space components of p_μ, viz., p_1, p_2, p_3, to be respectively the components p_x, p_y, p_z of the 3-momentum vector **p** for all velocities, then we must interpret $m_0/(1 - V^2/c^2)^{1/2}$ as the mass of the particle when it is moving with a velocity **V**. When the particle is at rest, **V** = 0 and its mass is m_0. This is known as the **rest mass** or the **proper mass** of the particle; it is the mass of the particle measured in an inertial frame of reference in which it is at rest. Hence according to the special theory of relativity, the mass of a particle is not a constant quantity. It depends upon the velocity of the particle and is given by the formula

$$m = \frac{m_0}{\sqrt{1 - V^2/c^2}}\ . \tag{3.10}$$

This relation shows that the mass m of a particle measured by an observer depends on its speed relative to the observer, i.e., on the reference frame from which it is measured, and is therefore an extrinsic property of the particle. In fact, the greater the speed of a particle relative to a given frame, the larger is its mass in that frame.

Equation (3.10) appears to be against everyday experience. In daily life one never observes a variation in mass with speed. This is because at ordinary speed the variation in mass with speed is extremely small. In fact, the relativistic increase in mass becomes significant only at speeds very close to that of light. Thus even at a tremendous speed such as 32000 km s^{-1}., the increase in mass is only about 0.5%. However, the mass is more than doubled when the speed is about 0.9c. It must be emphasized that relation (3.10) cannot be deduced logically; it is only a consequence of the way we relate the space components of p_μ with p_x, p_y, p_z. However, since its first verification made in 1908 by Bucherer by using electrons from a radioactive source, it has been confirmed experimentally a number of times.

With the aforementioned interpretation of mass, we may write

$$p_\mu = (\mathbf{p}, imc). \tag{3.11}$$

Accordingly, the 3-momentum and ic times the mass form the four components of 4-momentum just as the space coordinates and ict form the four components of the position 4-vector. Similar results are met throughout the special theory of relativity. For instance, force and power, and electric current density vector and charge density, occur together as 4-vectors. This unification of different quantities as the components of a 4-vector is a great beauty of the 4-dimensional approach to the theory.

We shall now derive an important relation between the momentum and mass of a particle. Consider two frames of reference such that in one of them the particle is at rest. Then the components of the 4-momentum in the two frames are

$$p_\mu = (p_1, p_2, p_3, p_4) = (p_x, p_y, p_z, imc) = (\mathbf{p}, imc)$$

$$p'_\mu = (p'_1, p'_2, p'_3, p'_4) = (0, 0, 0, im_0c) = (0, im_0c). \tag{3.12}$$

Since the magnitude of a 4-vector must remain invariant, we must have

$$p^2 - m^2c^2 = -m_0^2c^2$$

or

$$p^2 + m_0^2c^2 = m^2c^2. \tag{3.13}$$

It is interesting to note that in a closed system, with no energy or matter entering or leaving, this relation connects the principle of conservation of energy with that of conservation of momentum. This is a relation which is frequently used in modern physics.

To find the transformation law for 4-momentum, we note that it is a 4-vector. Therefore, we must have

$$p'_1 = \gamma\left(p_1 + i\frac{v}{c}p_4\right) \tag{3.14a}$$

$$p'_2 = p_2 \tag{3.14b}$$

$$p'_3 = p_3 \tag{3.14c}$$

$$p'_4 = \gamma\,(p_4 - i\,\frac{v}{c}\,p_1). \tag{3.14d}$$

But $$p_1 = \frac{m_0}{\sqrt{1 - V^2/c^2}}\,V_x = m\,V_x = p_x\,,$$

$$p_2 = p_y, \qquad p_3 = p_z$$

and $$p_4 = imc. \tag{3.11$'$}$$

Substituting these and similar expressions for primed quantities in equations (3.14), we get

$$p'_x = \gamma\,(p_x - m\,v) \tag{3.15a}$$

$$p'_y = p_y \tag{3.15b}$$

$$p'_z = p_z \tag{3.15c}$$

$$m' = \gamma\,(m - \frac{v}{c^2}\,p_x). \tag{3.15d}$$

The last equation gives us the transformation law for the mass of a particle.

4-Force and Equation of Motion

In classical mechanics, Newton's second law of motion may be taken as giving the definition of force. That is, if the momentum **p** of a particle changes, then the rate of change of momentum of the particle, d**p**/dt, is called the force **F** on the particle:

$$\mathbf{F} = \frac{d\mathbf{p}}{dt}$$

or $$F_i = \frac{dp_i}{dt}, \quad i = 1, 2, 3.$$

In special relativity, we are free to choose any definition of force which reduces to the classical expression for classical situations. From a mathematical point of view, it is convenient to introduce the concept of force by defining a 4-vector F_μ, called 4-force, as the rate of change of 4-momentum with respect to proper time:

$$F_\mu = \frac{dp_\mu}{d\tau}. \tag{3.16}$$

To check whether the space components of F_μ reduce to classical expressions for classical situations, we first write down the space components of F_μ explicitly. We have

$$F_1 = \frac{dp_1}{d\tau} = \frac{dp_1}{dt}\frac{dt}{d\tau}$$

$$= \frac{1}{\sqrt{1 - V^2/c^2}} \frac{d}{dt}(m\ V_x)$$

$$= m_0\ \gamma(V) \frac{d}{dt}[\gamma(V)\ V_x]. \tag{3.17a}$$

Similarly, we can obtain

$$F_2 = m_0\ \gamma(V) \frac{d}{dt}[\gamma(V)\ V_y] \tag{3.17b}$$

$$F_3 = m_0\ \gamma(V) \frac{d}{dt}[\gamma(V)\ V_z]. \tag{3.17c}$$

It may be stressed that the velocity **V** of a particle is, in general, a variable quantity. We note that for $V \ll c$, the spatial components F_1, F_2, F_3 of F_μ reduce to the classical expressions for the components of force, thus vindicating our definition of 4-force. Moreover, as both sides of equation (3.16) transform as 4-vectors, this equation is covariant under a Lorentz transformation.

Hence equations (3.17 a,b,c) can be taken as expressing the correct law of motion in relativistic mechanics. However, 3-force is no more an absolute quantity.

The fourth component of F_μ is given by

$$F_4 = \frac{dp_4}{d\tau} = \frac{dp_4}{dt}\frac{dt}{d\tau} = \gamma(V) \frac{d}{dt}(imc)$$

$$= ic\ \gamma(V) \frac{dm}{dt}. \tag{3.17d}$$

When we are considering a system of particles in special relativity, a difficulty arises in using equation (3.16) because we have to use a different proper time for each particle. A different definition of force, which is more convenient when we are considering a system of particles, is given through the relation

$$f_\mu = \frac{dp_\mu}{dt} . \tag{3.18}$$

It is this definition of force which is usually used in the literature. From the above relation, we have

$$f_1 = \frac{dp_1}{dt} = \frac{d}{dt}(m\, V_x). \tag{3.19a}$$

Since m is the mass of the particle when it is moving with a velocity **V**, the expression mV_x may be considered as the x-component of the 3-momentum provided we interpret m as mass of the particle when it is moving with a velocity **V**. Then the x-component of the 3-force may be written as

$$f_x = \frac{d}{dt}(m\, V_x).$$

Thus we obtain

$$f_1 = f_x.$$

Similarly, we can show that f_2 and f_3 are equal to f_y and f_z respectively, so that we may write

$$\mathbf{f} = \frac{d}{dt}(m\, \mathbf{V}). \tag{3.19b}$$

Moreover, $$f_4 = \frac{dp_4}{dt} = \frac{d}{dt}(imc) = ic\frac{dm}{dt} . \tag{3.19c}$$

Thus:

$$f_\mu = (f_1, f_2, f_3, f_4) = (f_x, f_y, f_z, ic\frac{dm}{dt}) = (\mathbf{f}, ic\frac{dm}{dt}). \tag{3.19d}$$

Problem

Prove that $\mathbf{f} \cdot \mathbf{V} = c^2 \frac{dm}{dt}$. (3.20)

Solution

We have shown that

$$V_\mu V_\mu = -c^2. \tag{3.3'}$$

Differentiating with respect to t, we get

$$V_\mu \frac{d}{dt}(V_\mu) = 0$$

or $$V_\mu \frac{d}{dt}(m_0 V_\mu) = 0$$

or $$V_\mu f_\mu = 0. \tag{3.21}$$

or $$V_1 f_1 + V_2 f_2 + V_3 f_3 + V_4 f_4 = 0$$

or $$\sum_{x,y,z} \gamma(V) V_x f_x - c^2 \gamma(V) \frac{dm}{dt} = 0,$$

where we have used equations (3.2) and the relations $f_{1,2,3} = f_{x,y,z}$.

or $$\mathbf{f} \cdot \mathbf{V} = c^2 \frac{dm}{dt} . \tag{3.20'}$$

This gives the required relation. Equation (3.19d) may then be written as

$$f_\mu = (\mathbf{f}, \frac{i}{c}\, \mathbf{f} \cdot \mathbf{V}). \tag{3.22}$$

Problem

Show that power is $-$ ic times the fourth component of f_μ.

It may be noted that

$$F_\mu = \frac{dp_\mu}{d\tau} = \frac{dp_\mu}{dt}\frac{dt}{d\tau} = \gamma(V)\, f_\mu. \tag{3.23}$$

This gives us the relation between two different definitions of force. It may be stressed that while F_μ is a 4-vector, f_μ is not. Therefore, although equation (3.18) can be taken as a definition of force, it cannot be the equation of motion in relativistic mechanics because it is not covariant under a Lorentz transformation.

The transformation law for $f_\mu = (f_x, f_y, f_z, f_4)$ may be obtained as follows:

Since F_μ is a 4-vector, in going from one inertial frame to another, its components will transform as

$$F_1' = \gamma\,(F_1 + i\,\frac{v}{c}\,F_4) \tag{3.24a}$$

$$F_2' = F_2 \tag{3.24b}$$

$$F_3' = F_3 \tag{3.24c}$$

$$F_4' = (F_4 - i\,\frac{v}{c}\,F_1). \tag{3.24d}$$

By using equations (3.23), (3.17a,d), (3.6) and a relation similar to equation (3.23) in the primed frame, equation (3.24a) can be written as

$$f_x' = A\,(f_x - v\,\frac{dm}{dt})$$

$$= A\,(f_x - \frac{v}{c^2}\,\mathbf{f}\cdot\mathbf{V}), \tag{3.25}$$

where we have used the result obtained in equation (3.21). From equations (3.6) and (3.25), we get

$$f_x' = (f_x - v\,\frac{dm}{dt})/(1 - \frac{v}{c^2}V_x)$$

$$= (f_x - \frac{v}{c^2}\mathbf{f}\cdot\mathbf{V})/(1 - \frac{v}{c^2}V_x). \tag{3.26a}$$

Similarly, we obtain

$$f_y' = f_y/[\gamma\,(1 - \frac{v}{c^2}V_x)] \tag{3.26b}$$

$$f_z' = f_z/[\gamma\,(1 - \frac{v}{c^2}V_x)]. \tag{3.26c}$$

In particular, if the velocity of the particle in the S-frame is parallel to the x-axis: $\mathbf{V} = (V, 0, 0)$, and the force is also acting in the same direction: $\mathbf{f} =$

(f, 0, 0), then the above equations give

$$f'_x = (f - v\frac{dm}{dt})/(1 - \frac{v}{c^2}V_x) \tag{3.27a}$$

$$f'_y = 0 \tag{3.27b}$$

$$f'_z = 0. \tag{3.27c}$$

This shows that in this case $\mathbf{f}'$, the 3-force in S′, will also be parallel to the common x-axis.

Problem

Show that the force **f** in the instantaneous rest frame of a particle is greater than the corresponding force in any other frame.

Substituting the expressions for F_1, F_4, F'_4, from equations (3.17a) and from a relation similar to equation (3.17d) in equation (3.24d) and making use of equation (3.23), we get

$$f'_4 = \frac{\gamma(V)}{\gamma(V')}(f_4 - i\frac{v}{c}f_x). \tag{3.28}$$

Now, as

$$\mathbf{f} = \frac{d}{dt}(m\,\mathbf{V}) = \frac{dm}{dt}\mathbf{V} + m\frac{d\mathbf{V}}{dt} = m\,\mathbf{a} + \frac{\mathbf{f}\cdot\mathbf{V}}{c^2}\mathbf{V}, \tag{3.29}$$

we find that, in general, the 3-acceleration **a** is not in the same direction as the force **f**. There are, however, two cases in which **f** and **a** are parallel to each other. Firstly, if the force **f** is perpendicular to V, then the second term vanishes and the acceleration is in the same direction as the force. Secondly, if **f** is parallel to V, then the above equation may be written as

$$f\hat{\mathbf{f}} = m\,\mathbf{a} + \frac{\mathbf{f}\cdot\mathbf{V}}{c^2}V\hat{\mathbf{f}},$$

where $\hat{\mathbf{f}}$ is a unit vector in the direction of the force,

or $$(f - \frac{\mathbf{f}\cdot\mathbf{V}}{c^2}V)\hat{\mathbf{f}} = m\,\mathbf{a}. \tag{3.30}$$

We find that in this case too, the acceleration and the force are in the same direction. Since **a** is parallel to both **f** and **V**, the particle moves in a straight line. The motion of a charged particle in a uniform electric field when its initial velocity is in the direction of the field is an example of such a case. When **f** and **a** are parallel to the velocity **V**, then by substituting

$$m = \frac{m_0}{\sqrt{1 - V^2/c^2}}$$

in equation

$$f = \frac{dm}{dt} V + m \frac{dV}{dt},$$

we get

$$f = \frac{m_0}{(1 - V^2/c^2)^{3/2}} a, \qquad \text{where} \qquad a = \frac{dV}{dt}.$$

The quantity $m_0/(1 - V^2/c^2)^{3/2}$ is sometimes called the **longitudinal mass.** Similarly, when **f** and **a** are perpendicular to the velocity **V**, we obtain

$$f = \frac{m_0}{\sqrt{1 - V^2/c^2}} a.$$

The quantity $m_0/(1 - V^2/c^2)^{1/2}$ is sometimes called the **transverse mass.**

Motion under a Constant Force

As a simple application of the preceding formulae, we consider the motion of a particle, under the action of a constant force **f**, with its initial velocity in the direction of the applied force. We have shown in the preceding article that in this case the acceleration **a** of the particle will also be along **f**, and, therefore, it will continue to move in the direction of the constant force. The path of the particle will therefore be a straight line which may be chosen as x-axis of a coordinate system. Then for this particle the equation of motion $d\mathbf{p}/dt = \mathbf{f}$ can be written as

$$\frac{dþ}{dt} = f,$$

where þ stands for the magnitude of 3-momentum **p**. Here f, being the magnitude of the constant applied force, is also a constant. Integrating with respect to t, we get

$$\text{þ} = f\,t + K,$$

where K is a constant. Let us assume that the particle is initially at rest, i.e., þ = 0 for t = 0. Then the above equation gives K = 0 and, therefore, reduces to

$$\text{þ} = f\,t$$

or

$$m\,V = f\,t$$

or

$$\frac{m_0}{\sqrt{1 - V^2/c^2}}\,V = f\,t.$$

Squaring and rearranging the terms, we get

$$V = \frac{dx}{dt} = \frac{a\,t}{\sqrt{1 + a^2t^2/c^2}} = a\,t\,(1 + a^2\,t^2/c^2)^{-1/2}, \tag{3.31}$$

where $a = f/m_0$ and is therefore a constant quantity.

Integrating equation (3.31) with respect to time, we get

$$x = \frac{c^2}{a}\sqrt{1 + a^2t^2/c^2} + K', \tag{3.32}$$

where K' is a constant. Assuming that at time t = 0, the particle is at the origin, i.e., at x = 0, we get $K' = -\,c/a$. Substituting this value in equation (3.32), we get

$$x + \frac{c^2}{a} = \frac{c^2}{a}\sqrt{1 + a^2t^2/c^2}\,. \tag{3.33}$$

Squaring and rearranging the terms, we obtain

$$(x + \frac{c^2}{a})^2 - c^2\,t^2 = \frac{c^4}{a}. \tag{3.34}$$

This is the equation of a hyperbola in the xt-plane. For this reason the motion of a particle moving under the action of a constant force, with its initial velocity in the direction of the force, is called the hyperbolic motion.

From equation (3.33), the position x of the particle at any time is given by

$$x = -\frac{c^2}{a} + \frac{c^2}{a}\sqrt{1 + a^2t^2/c^2}\,.$$

But for small values of time, $at \ll c$, and, therefore, by binomial theorem, the above equation reduces to

$$x \approx -\frac{c^2}{a} + \frac{c^2}{a}\left(1 + \frac{1}{2}\frac{a^2\,t^2}{c^2}\right)$$

or
$$x \approx \frac{1}{2}\,a\,t^2, \tag{3.35}$$

which is the classical result.

If a particle of charge e moves under the influence of a constant (in time) and uniform (in space) longitudinal electric field ϵ, then $a = e\epsilon/m_0$ and equation (3.33) yields

$$x = \frac{m_0\,c^2}{e\,\epsilon}\left[\left(1 + \frac{e^2\,\epsilon^2\,t^2}{m_0^2\,c^2}\right)^{1/2} - 1\,\right]. \tag{3.36}$$

This equation gives the position of the charged particle at any time t.

Mass-Energy Relation

The most significant relation obtained from the special theory of relativity concerns mass and energy. As in classical mechanics, the kinetic energy of a moving body is defined as equal to the work done on the body in bringing it to that state of motion from its state of rest. Therefore, the kinetic energy dT acquired by the body when it is displaced by a force **f** through a distance d**r** is given by

$$dT = \mathbf{f} \cdot d\mathbf{r}, \tag{3.37}$$

so that the rate of increase of kinetic energy is

$$\frac{dT}{dt} = \mathbf{f} \cdot \frac{d\mathbf{r}}{dt} = \mathbf{f} \cdot \mathbf{V},$$

where **V** is the velocity of the particle.

or
$$\frac{dT}{dt} = \left[\frac{d}{dt}(m\,\mathbf{V})\right] \cdot \mathbf{V} = \frac{dm}{dt}\mathbf{V} \cdot \mathbf{V} + m\,\frac{d\mathbf{V}}{dt} \cdot \mathbf{V}$$

$$= V^2 \frac{dm}{dt} + m\,V \frac{dV}{dt}. \tag{3.38}$$

But $$m = \frac{m_0}{\sqrt{1 - V^2/c^2}}.$$

Differentiating m with respect to t, we get

$$\frac{dm}{dt} = \frac{m_0}{(1 - V^2/c^2)^{3/2}} V \frac{dV}{dt} = \frac{m\,V}{c^2(1 - V^2/c^2)} \frac{dV}{dt}$$

or $$\frac{dV}{dt} = \frac{c^2\,(1 - V^2/c^2)}{m\,V} \frac{dm}{dt}.$$

Substituting this expression for dV/dt in equation (3.38), we get

$$\frac{dT}{dt} = V^2 \frac{dm}{dt} + c^2\,(1 - V^2/c^2) \frac{dm}{dt} = c^2 \frac{dm}{dt}. \tag{3.39}$$

Integrating with respect to t, we get

$$T = m\,c^2 + K,$$

where K is a constant

or $$T = \frac{m_0}{\sqrt{1 - V^2/c^2}} c^2 + K.$$

If the kinetic energy of the particle at rest is taken as zero, then K must be equal to $- m_0\,c^2$. Hence

$$T = m\,c^2 - m_0\,c^2 = \frac{m_0}{\sqrt{1 - V^2/c^2}} c^2 - m_0\,c^2. \tag{3.40}$$

This gives us an expression for the kinetic energy of the particle. For $V \ll c$, expanding $(1 - V^2/c^2)^{-1/2}$ by binomial theorem, we obtain

$$T = m_0\,c^2\,(1 + \frac{1}{2} V^2/c^2 + \frac{3}{8} V^4/c^4 + \cdots) - m_0\,c^2$$

$$\approx \frac{1}{2} m_0 V^2,$$

where V^4/c^4 and higher order terms in V/c have been neglected. This is the classical formula for the kinetic energy of a particle. Note that the formula $T = m_0 V^2/2$ is valid only for $V \ll c$.

Since m_0c^2 is a constant, the equation $T = mc^2 - m_0c^2$ shows that any increase in the kinetic energy of a particle will cause an increase in its mass. Einstein made the bold assumption that not only this additional mass but all mass is a manifestation of energy, i.e., m_0c^2 may be considered as the intrinsic energy of the particle when it is at rest while $mc^2 = m_0c^2 + T$ has the physical significance of the total energy E of a free particle:

$$E = mc^2. \tag{3.41}$$

The quantity m_0c^2 is called the **rest mass energy** of the particle. Thus according to the hypothesis of Einstein, the formula $E = mc^2$ expresses the fact that mass and energy are interconvertible such that if a mass m is changed into energy, the amount of energy produced is mc^2. There exist innumerable examples, including the release of fantastic amounts of energy by atomic and hydrogen bombs and the power production by nuclear reactors, which have vindicated this bold hypothesis.

The above discussion shows that energy E should possess all the properties, such as inertia, of mass E/c^2.

Problem

Show that 1 kilogram of matter when annihilated would produce 9×10^6 joules of energy.

Significance of the Fourth Component of 4-Momentum

We have proved that

$$p_4 = imc. \tag{3.11$'$}$$

Substituting the expression for m from the relation $E = mc^2$ in equation (3.11′), we get

$$p_4 = \frac{i}{c}E. \tag{3.42}$$

Thus apart from the constant factor i/c, energy is the fourth component of 4-momentum.

Problem

Find the square of the magnitude of 4-momentum. Show that under Lorentz transformation without rotations, the transformation equations for momentum and energy are given by

$$\mathbf{p} = \mathbf{p}' + \frac{\mathbf{v}\,[\,\mathbf{v}\cdot\mathbf{p}'\,\{1-\sqrt{1-v^2/c^2}\,\} + E'\,v^2/c^2]}{v^2\sqrt{1-v^2/c^2}}\,, \tag{3.43a}$$

$$E = \frac{E' + \mathbf{v}\cdot\mathbf{p}'}{\sqrt{1-v^2/c^2}}\,. \tag{3.43b}$$

Again, multiplying equation (3.10) by c^2, we obtain

$$m'c^2 = \gamma\,(mc^2 - v\,p_x)$$

or

$$E' = \gamma\,(E - v\,p_x). \tag{3.44}$$

This is the transformation law for the total energy.

Now, p^2, the square of the magnitude of the momentum 4–vector ($\mathbf{p}$, iE/c), is given by the relation

$$p^2 = \mathbf{p}^2 - E^2/c^2.$$

But, from equations (3.13) and (3.41), we have

$$E^2 = m_0^{\,2}c^4 + c^2\,\mathbf{p}^2, \tag{3.45}$$

so that equation (3.45) reduces to $p^2 = -\,m_0^{\,2}c^2$. That is, the rest mass, which is invariant and therefore has the same value in every inertial system, is a measure of the magnitude of the momentum 4-vector. On the other hand, energy behaves as the time component of the same 4-vector and is not an invariant quantity; it will not have the same value in every inertial frame.

We find from equation (3.40) that for $V \to c$, the kinetic energy T of a particle tends to infinity so that only an infinite amount of work can accelerate a particle up to the velocity of light. Since an infinite amount of work cannot be done in finite time, no particle with non-vanishing rest mass can attain the velocity of light which therefore plays the role of a limiting velocity. Later on we shall show that particles of zero rest mass, called photons, can have a velocity equal to that of light.

An important consequence of this fact is the non-existence of rigid bodies. In classical mechanics, a rigid body is defined to be a system of particles such that the distance between them always remains the same. Such a body does not exist in nature because on the application of a sufficiently large force every real object will bend or stretch and consequently be deformed to some extent. Special relativity, however, tells us that a rigid body cannot exist even in principle. In order to see this, suppose that an extended body is set into motion by an external force applied at one of its points. If this body is rigid, then, in order that it may not be deformed, all its points should start moving at the same moment at which the external force is applied at one of its points. This is possible in classical mechanics since information can be communicated at infinite speed. This is impossible in special relativity where forces can be transmitted from one point of a body to another only at a finite speed. Hence, on the application of force, an extended body will always be deformed and, for that reason, cannot be considered as a rigid body. We, therefore, come to the conclusion that *in special relativity the existence of a rigid body is an impossibility*.

The impossibility of the existence of a rigid body leads to an interesting result regarding elementary particles. An elementary particle is defined as an object which participates in all physical phenomena as a unit. If it possesses finite dimensions, i.e., it is extended in space, then according to special relativity it would be deformable as rigid bodies cannot be realized even in principle. Such a deformation will refute the existence of an elementary particle as a unit. We conclude that an elementary particle cannot have finite dimensions and therefore must be regarded as a geometrical point.

Another interesting point which deserves a careful consideration is the following. The entire structure of special relativity is based on inertial frames of reference which are supposed to be rigid bodies. Since according to the special relativity, the existence of a rigid body is impossible, the entire structure of relativity must collapse. The answer to this dilemma lies in the fact that if we do not apply forces on our frames of reference, we shall not encounter this problem because in the absence of forces the concept of rigidity would not arise. Since, by definition, an inertial frame is unaccelerated, this condition is automatically satisfied. However, a reference frame will always interact with the material system under observation. This cannot be avoided. Moreover, the very act of observation and communication will also cause interaction between the observer and the frame. This last objection can be overcome by assuming that such interactions are negligible, i.e., the system and the frame of reference are very heavy as compared with the objects carrying the information.

Examples of Equivalence of Mass and Energy

We shall now consider some important examples of the equivalence of mass and energy which have been confirmed experimentally.

Binding Energy

We know that a nucleus is composed of protons and neutrons. It may therefore be expected that the mass of a nucleus would be equal to the sum of the masses of its constituent particles. A survey of nuclear mass tables, however, shows that this is not true. The mass of a stable nucleus is always less than the sum of the masses of its constituent particles. This difference in mass is easily accounted for by assuming that the mass difference Δm is changed into energy ΔE ($\Delta E = c^2 \Delta m$) which is used to bind the nucleons together. This is called the binding energy of the nucleus and is evidently equal to the energy required to break up the nucleus into its constituent particles. We shall illustrate this by an example. Helium nucleus, ${}^4_2He$, consists of two protons and two neutrons and its mass is 4.003874 amu, where 1 amu is 1/12 of the mass of ${}^{12}_6C$ and is equal to 1.66033×10^{-27} kg. The mass of a proton is 1.007646 amu while that of a neutron is 1.009034 amu. Therefore

the mass of (2 protons + 2 neutrons)

$= (2 \times 1.007646 + 2 \times 1.009034)$ amu

$= 4.03336$ amu.

This is greater than the mass of the helium nucleus; the mass difference being

0.02949 amu. The nucleus is therefore stable, as an energy of 27.46 MeV, equivalent to the above mass difference, is required to break it into its constituent particles. This is the binding energy of the helium nucleus.

Problem

Show that the binding energy of the deuteron nucleus is 2.226 MeV.

Natural Radioactivity

Some radioactive nuclides disintegrate spontaneously, say, by the emission of α-particles. It is found that for such substances, the sum of the masses of the products of disintegration is less than the mass of the disintegrating nuclide. The corresponding difference in mass is released in the form of energy. We shall illustrate this by considering the disintegration of radium into radon and helium:

$$^{226}_{88}\mathrm{Ra} \rightarrow {}^{222}_{86}\mathrm{Rn} + {}^{4}_{2}\mathrm{He}.$$

The mass of $^{226}_{88}\mathrm{Ra}$ nucleus (226.09600 amu) is greater than the sum of the masses of nuclei of $^{222}_{86}\mathrm{Rn}$ (222.08690 amu) and $^{4}_{2}\mathrm{He}$ (4.003874 amu) by 0.00523 amu. Hence radium is unstable against α-decay. The energy released in disintegration is c times the mass difference so that $E = \Delta m\ c^2 = 4.87$ MeV.

Nuclear Reactions

Energy may also be released in nuclear reactions which are particularly suited for verifying the law $E = mc^2$. The nuclear masses are known with great precision and, more important, the velocities of the final particles are large enough so that the diminution in mass can be appreciable. For instance, consider the reaction

$$^{7}_{3}\mathrm{Li} + {}^{1}_{1}\mathrm{H} \rightarrow 2\,{}^{4}_{2}\mathrm{He}$$

in which two particles ($^{4}_{2}\mathrm{He}$ nuclei) per reaction are produced when lithium ($^{7}_{3}\mathrm{Li}$) is bombarded with monoenergetic protons. We have

$$\text{Mass of } {}^{7}_{3}\mathrm{Li} = 7.018232 \text{ amu}$$

Mass of ${}^{1}_{1}H$ = 1.007646 amu

Mass of ${}^{4}_{2}He$ = 4.003874 amu.

Therefore, the difference in mass

$$\Delta m = (7.018232 + 1.007646) - 2 \times 4.003874 \text{ amu}$$

$$= 0.011813 \text{ amu.}$$

This corresponds to an energy of 16.88 MeV. If T is the kinetic energy of the incident proton, then the total energy released in the nuclear reaction should be (16.88 + T) MeV and would appear as the kinetic energy of α-particles.

Fission

It is found that if ${}^{235}U$ nucleus is bombarded with slow neutrons, it usually splits up into two lighter nuclei of nearly equal mass. In addition to that, some neutrons are emitted and energy is released. This process is known as **fission**. The sum of the masses of the nuclei produced and the neutrons released in fission is less than the mass of the uranium nucleus. This difference of mass appears as an enormous release of energy. It is found that about 200 MeV of energy is released per nuclear fission. It may be noted that fission is not confined to ${}^{235}U$ nuclei. It can also take place in other heavy elements with release of large amounts of energy. The principle of nuclear fission has been made use of in the construction of atomic bombs and nuclear reactors.

Pair Production

When a high energy photon passes in the vicinity of matter, sometimes it destroys itself to create an electron and a positron. This process is called electron-positron pair-production. It is the materialization of energy. The rest mass of an electron or a positron is equivalent to 0.51 MeV. Thus to create an electron-positron pair, the photon energy must be greater than the rest-mass of the electron-positron pair, i.e., 1.02 MeV. It should be noted that a photon cannot produce a pair in free-space because then the laws of conservation of energy and momentum cannot hold simultaneously. The presence of an atomic nucleus in the vicinity of the photon is essential for the creation of a pair. The nucleus, because of its preponderant mass, acquires negligible energy but obtains sufficient momentum so as to enable the system to satisfy the laws of conservation of energy and momentum.

The reverse of the process can also occur. When an electron and a positron come close together, they annihilate each other producing radiation. This is known as pair-annihilation. Usually two photons per pair-annihilation are produced. Pair-creation and pair-annihilation are not limited to electrons and positrons. Other particle-antiparticle pairs can also be created and annihilated.

Stellar Energy

The sun has been radiating enormous energy for billions of years without showing any signs of getting cold. Where does this energy come from? It is now believed that the most important source of energy of the sun is the nuclear burning of protons to create helium according to the following cycle of reactions:

$$ {}^{1}_{1}\mathrm{H} + \mathrm{p} \rightarrow {}^{2}_{1}\mathrm{H} + e^{+} + \nu $$

$$ {}^{2}_{1}\mathrm{H} + \mathrm{p} \rightarrow {}^{3}_{2}\mathrm{He} + \gamma $$

$$ {}^{3}_{2}\mathrm{He} + {}^{3}_{2}\mathrm{He} \rightarrow {}^{4}_{2}\mathrm{He} + 2\,{}^{1}_{1}\mathrm{H}. $$

This series is known as the proton-proton cycle. It produces helium at the expense of hydrogen. Since the mass of four protons (hydrogen nuclei) which fuse together to form a helium nucleus is less than the mass of the nucleus itself, energy is released during this process which is known as **fusion.** It is the tremendous amount of energy produced in this manner which keeps the sun shining.

Problem

Calculate the amount of energy released per nucleus during the fusion described by the above relations.

Motion of a Charged Particle in a Uniform Transverse Electric Field

Let us next find the trajectory of a particle of charge e and rest mass m_0 moving in a *uniform* transverse electric field of magnitude ϵ, i.e., in a uniform electric field *acting at right angles* to the initial direction of motion of the particle. Suppose that the particle is projected with initial momentum p_0. Take the point of projection of the charged particle as the origin of a coordinate system and the direction of projection of the particle as positive x-axis and suppose that the uniform electric field is acting along the positive y-axis. If **i, j, k** are unit vectors along the x, y, z-axes respectively and **p** is the 3-momentum of the particle at any instant, then the equation of motion for the particle may be written as

$$\frac{d\mathbf{p}}{dt} = e\,\epsilon\,\mathbf{j}. \tag{3.46}$$

Equating the coefficients of **i, j, k** on the two sides of this equation, we get

$$\frac{dp_x}{dt} = 0, \quad \frac{dp_y}{dt} = e\,\epsilon, \quad \frac{dp_z}{dt} = 0. \tag{3.47}$$

Integrating with respect to t, we obtain

$$p_x = A, \quad p_y = e\,\epsilon\,t + B, \quad p_z = D, \tag{3.48}$$

where A, B and D are constants. Since the particle is projected from the origin with momentum p_0 along the x-axis at $t = 0$, we have

$$p_x = p_0, \quad p_y = 0, \quad p_z = 0.$$

Therefore, $A = p_0$, $B = D = 0$ and equations (3.48) become

$$p_x = p_0, \quad p_y = e\,\epsilon\,t, \quad p_z = 0. \tag{3.49}$$

or

$$m\,V_x = p_0, \quad m\,V_y = e\,\epsilon\,t, \quad m\,V_z = 0. \tag{3.50}$$

or

$$\frac{dx}{dt} = p_0/m, \quad \frac{dy}{dt} = e\epsilon t/m, \quad \frac{dz}{dt} = 0,$$

where m is the mass of the particle when it is moving with a velocity **V**.

or

$$\frac{dx}{dt} = p_0\,c^2/E, \quad \frac{dy}{dt} = e\epsilon tc^2/E, \quad \frac{dz}{dt} = 0. \tag{3.51}$$

Now, by virtue of equation (3.45), the total energy E of the particle is given by

$$E^2 = m_0^2 c^4 + p^2 c^2 = m_0^2 c^4 + (p_x^2 + p_y^2 + p_z^2) c^2. \tag{3.52}$$

Substituting the expressions for p_x, p_y, p_z from equation (3.49) in equation (3.52), we get

$$E^2 = m_0^2 c^4 + p_0^2 c^2 + e^2 \epsilon^2 t^2 c^2$$

$$= E_0^2 + e^2 \epsilon^2 t^2 c^2, \tag{3.53}$$

where E_0 is the initial energy of the particle when its 3-momentum is $\mathbf{p}_0$. Substituting the expression for E from the above equation in equation (3.51), we get

$$\frac{dx}{dt} = \frac{p_0 c^2}{\sqrt{E_0^2 + e^2 \epsilon^2 t^2 c^2}}$$

$$\frac{dy}{dt} = \frac{e \epsilon t c^2}{\sqrt{E_0^2 + e^2 \epsilon^2 t^2 c^2}}.$$

Since the electric field ϵ is uniform, integrating with respect to t, we get

$$x = \frac{p_0 c}{e \epsilon} \sinh^{-1} \left(\frac{e \epsilon t c}{E_0} \right) + K \tag{3.54}$$

and

$$y = \frac{1}{e \epsilon} (E_0^2 + e^2 \epsilon^2 t^2 c^2)^{1/2} + K'. \tag{3.55}$$

For $t = 0$, $x = 0$, $y = 0$. This yields $K = 0$ and $K' = - E_0/e \epsilon$. Substituting these values in equations (3.54) and (3.55), we obtain

$$x = \frac{p_0 c}{e \epsilon} \sinh^{-1} \left(\frac{e \epsilon t c}{E_0} \right) \tag{3.56}$$

$$y = \frac{1}{e \epsilon} (E_0^2 + e^2 \epsilon^2 t^2 c^2)^{1/2} - E_0/e \epsilon. \tag{3.57}$$

Eliminating t between equations (3.56) and (3.57), we have

$$y = \frac{1}{e \epsilon} \left[\left\{ E_0^2 + e^2 \epsilon^2 c^2 \left(\frac{E_0}{e c \epsilon} \right)^2 \sinh^2 \left(\frac{e \epsilon x}{p_0 c} \right) \right\}^{1/2} - E_0 \right]$$

$$= \frac{1}{e \epsilon} [\{ E_0^2 + E_0^2 \sinh^2 \left(\frac{e \epsilon x}{p_0 c} \right) \}^{1/2} - E_0]$$

$$= \frac{1}{e\,\epsilon} [\{ E_0^2 \cosh^2 (\frac{e\,\epsilon\,x}{p_0 c}) \}^{1/2} - E_0]$$

$$= \frac{E_0}{e\,\epsilon} [\cosh (\frac{e\,\epsilon\,x}{p_0 c}) - 1]. \qquad (3.58)$$

This is the equation of the path of a charged particle moving in a uniform transverse electric field.

Problem

Find the angle ϕ through which the charged particle projected along the x-axis is deflected by the uniform transverse electric field.

Motion of a Charged Particle in a Magnetic Field

Let us consider the motion of a particle of charge e projected in a magnetic field **H**. It will be shown in Chapter 5 that the force **f** acting on this particle at any instant is given by

$$\mathbf{f} = \frac{e}{c} \mathbf{V} \wedge \mathbf{H}, \qquad (3.59)$$

where **V** is the velocity of the particle. Since force is equal to the rate of change of momentum, we have

$$\frac{d\mathbf{p}}{dt} = \frac{e}{c} \mathbf{V} \wedge \mathbf{H}. \qquad (3.60)$$

Multiplying both sides scalarly by **p**, we get

$$\mathbf{p} \cdot \frac{d\mathbf{p}}{dt} = \frac{e}{c} \mathbf{p} \cdot \mathbf{V} \wedge \mathbf{H} = 0,$$

because in the triple product momentum and velocity are in the same direction.

or $$\frac{d\mathbf{p}^2}{dt} = 0.$$

Integrating with respect to t, we get

$$\mathbf{p}^2 = \text{þ}^2 = \text{constant}$$

or $$\text{þ} = \text{constant}, \tag{3.61}$$

i.e., the magnitude of the 3-momentum of a charged particle moving under the influence of a pure magnetic field is constant. This leads to the result that the speed of the charged particle is also constant. Thus if V is the initial speed of the particle, then its speed throughout the motion will remain V.

Now, as þ is constant, we may write

$$\frac{d\mathbf{p}}{dt} = \frac{d}{dt}(\text{þ}\ \mathbf{t}) = \text{þ}\ \frac{d\mathbf{t}}{dt},$$

where $\mathbf{t}$ a unit vector in the direction of motion of the particle,

or $$\frac{d\mathbf{p}}{dt} = \text{þ}\ \frac{d\mathbf{t}}{ds}\ \frac{ds}{dt} = \text{þ}\ \frac{V}{\rho}\ \mathbf{n}, \tag{3.62}$$

where s is the distance measured along the trajectory from some fixed point on it, ρ is the radius of curvature of the trajectory at the position of the particle, $\mathbf{n}$ is a unit vector along the principal normal of the trajectory thereat, ds/dt is the speed of the particle, and we have made use of the fact that $d\mathbf{t}/dt = \mathbf{n}/\rho$. Substituting this expression for $d\mathbf{p}/dt$ in equation (3.60), we get

$$\text{þ}\ \frac{V}{\rho}\ \mathbf{n} = \frac{e}{c}\ \mathbf{V} \wedge \mathbf{H}.$$

or $$\text{þ}\ \frac{V}{\rho} = \frac{e}{c}\ V\ H \sin\theta,$$

where θ is the angle between $\mathbf{V}$ and $\mathbf{H}$. The radius of curvature of the trajectory is therefore given by

$$\rho = \frac{c\ \text{þ}}{e\ H \sin\theta}. \tag{3.63}$$

We shall next show that if the magnetic field $\mathbf{H}$ is uniform, then the angle θ is constant.

Multiplying both sides of the equation

$$\frac{d\mathbf{p}}{dt} = \frac{e}{c}\ \mathbf{V} \wedge \mathbf{H} \tag{3.60$'$}$$

scalarly by $\mathbf{H}$, we get

$$\mathbf{H} \cdot \frac{d\mathbf{p}}{dt} = 0$$

or $$\mathbf{H} \cdot \frac{d(þ\mathbf{t})}{dt} = 0,$$

or $$\mathbf{H} \cdot \frac{d\mathbf{t}}{dt} = 0,$$

because þ is constant in this case. For a *uniform* magnetic field, we can write this equation as

$$\frac{d}{dt}(\mathbf{H} \cdot \mathbf{t}) = 0 \,.$$

Integrating with respect to time, we get

$$\mathbf{H} \cdot \mathbf{t} = \text{const.}$$

or $$H \cos\theta = \text{const.}$$

or $$\theta = \text{const.},$$

because H is uniform. Thus, in this case, all the quantities on the right hand side of equation (3.63) are constant. Therefore ρ is also constant so that the path of the charged particle in a uniform magnetic field is a circular helix.

If the charged particle is initially projected in a direction which is perpendicular to **H**, its velocity will always remain perpendicular to it. Therefore, for $\theta = 90^\circ$, equation (3.63) reduces to

$$\rho = \frac{c\,þ}{e\,H} = \frac{c\,m\,V}{e\,H}$$

or $$m\frac{V}{r} = \frac{e}{c}H, \tag{3.64}$$

where m is the relativistic mass of the charged particle and, following the usual practice, we have written r for ρ. This shows that the path of a charged particle in a uniform magnetic field is a circle provided that the velocity of the particle is initially perpendicular to **H**. Equation (3.64) also shows that the radius r of the circular motion is proportional to the momentum of the charged particle. The time T taken by the charged particle to complete one circle is given by

$$T = \frac{2\pi}{\omega} = 2\pi \frac{r}{V} = 2\pi \frac{m\,c}{e\,H}, \tag{3.65}$$

where ω is the angular velocity and where we have used equation (3.64). Thus the relativistic time period depends upon the mass of the particle. This fact is of great importance in contemporary physics and technology, and has been made use of in the construction of high-energy accelerators such as synchrotron and synchrocyclotron.

The Relativistic Lagrangian and Hamiltonian of a Charged Particle Moving in an Electromagnetic Field

Let us now determine the relativistic expressions for the Lagrangian and the Hamiltonian of a charged particle of rest mass m_0 moving with a velocity $\mathbf{V}\ (\equiv \frac{d\mathbf{r}}{dt} \equiv \dot{\mathbf{r}})$ in an electromagnetic field represented by the vector and scalar potentials $\mathbf{A}$ and ϕ respectively. It is well known that the force $\mathbf{F}$ acting on a particle of charge e moving with a velocity $\mathbf{V}$ in an electromagnetic field is given by

$$\mathbf{F} = e\,\mathbf{E} + \frac{e}{c}\,\mathbf{V} \wedge \mathbf{H}. \tag{3.66}$$

But the electric and magnetic fields are given by

$$\mathbf{E} = -\,\mathrm{grad}\,\phi - \frac{1}{c}\frac{\partial \mathbf{A}}{\partial t}$$

$$\mathbf{H} = \mathrm{curl}\,\mathbf{A},$$

while the force $\mathbf{F}$ can be written as

$$\mathbf{F} = \frac{dp}{dt} = \frac{d}{dt}(m\mathbf{V}) = \frac{d}{dt}\left(\frac{m_0}{\sqrt{1 - V^2/c^2}}\,\mathbf{V}\right)$$

so that equation (3.66) takes the form

$$\frac{d}{dt}\left(\frac{m_0}{\sqrt{1 - V^2/c^2}}\,\mathbf{V}\right) = -\,e\,\mathrm{grad}\,\phi - \frac{e}{c}\frac{\partial \mathbf{A}}{\partial t} + \frac{e}{c}\,\mathbf{V} \wedge \mathrm{curl}\,\mathbf{A}. \tag{3.67}$$

The x-component of this equation is

$$\frac{d}{dt}\left[\frac{m_0}{\sqrt{1-V^2/c^2}}V_x\right] = -e\frac{\partial\phi}{\partial x} - \frac{e}{c}\frac{\partial A_x}{\partial t}$$

$$+\frac{e}{c}[V_y(\frac{\partial A_y}{\partial x} - \frac{\partial A_x}{\partial y}) - V_z(\frac{\partial A_x}{\partial z} - \frac{\partial A_z}{\partial x})],$$

which may be written as

$$\frac{d}{dt}\left[m_0\,\dot{x}/\{1-(\dot{x}^2+\dot{y}^2+\dot{z}^2)/c^2\}^{1/2}\right]$$

$$= -e\frac{\partial\phi}{\partial x} - \frac{e}{c}\frac{\partial A_x}{\partial t} + \frac{e}{c}[\dot{y}(\frac{\partial A_y}{\partial x} - \frac{\partial A_x}{\partial y}) - \dot{z}(\frac{\partial A_x}{\partial z} - \frac{\partial A_z}{\partial x})], \quad (3.68)$$

where $\dot{x}$, $\dot{y}$ and $\dot{z}$ stand for $\frac{dx}{dt}$, $\frac{dy}{dt}$ and $\frac{dz}{dt}$, respectively.

Since $\frac{dA_x}{dt} = \frac{\partial A_x}{\partial t} + \dot{x}\frac{\partial A_x}{\partial x} + \dot{y}\frac{\partial A_x}{\partial y} + \dot{z}\frac{\partial A_x}{\partial z}$,

equation (3.68) may be written as

$$\frac{d}{dt}\left[m_0\,\dot{x}/\{1-(\dot{x}^2+\dot{y}^2+\dot{z}^2)/c^2\}^{1/2}\right]$$

$$= -e\frac{\partial\phi}{\partial x} - \frac{e}{c}[\frac{dA_x}{dt} + \dot{x}\frac{\partial A_x}{\partial x} + \dot{y}\frac{\partial A_x}{\partial y} + \dot{z}\frac{\partial A_x}{\partial z}] + \frac{e}{c}[\dot{y}\frac{\partial A_y}{\partial x} - \dot{y}\frac{\partial A_x}{\partial y}$$

$$-\dot{z}\frac{\partial A_x}{\partial z} + \dot{z}\frac{\partial A_z}{\partial x}]$$

or $$\frac{d}{dt}\left[m_0\,\dot{x}/\{1-(\dot{x}^2+\dot{y}^2+\dot{z}^2)/c^2\}^{1/2} + \frac{e}{c}A_x\right]$$

$$= -e\frac{\partial\phi}{\partial x} + \frac{e}{c}[\dot{x}\frac{\partial A_x}{\partial x} + \dot{y}\frac{\partial A_y}{\partial x} + \dot{z}\frac{\partial A_z}{\partial x}]$$

or $$\frac{d}{dt}\frac{\partial}{\partial\dot{x}}\left[-m_0c^2\{1-(\dot{x}^2+\dot{y}^2+\dot{z}^2)/c^2\}^{1/2} + \frac{e}{c}\dot{x}A_x\right]$$

$$= \frac{\partial}{\partial x}[-e\phi + \frac{e}{c}[\dot{x}A_x + \dot{y}A_y + \dot{z}A_z]$$

$$= \frac{\partial}{\partial x}[-e\phi + \frac{e}{c}\mathbf{V}\cdot\mathbf{A}]. \tag{3.69}$$

We can add the terms $\frac{e}{c}\dot{y}A_y$, $\frac{e}{c}\dot{z}A_z$, $-e\phi$ to the expression in square brackets on the left hand side of equation (3.69) without affecting its value because the derivatives of these terms with respect to $\dot{x}$ vanish. Similarly, we can add

$$\frac{\partial}{\partial x}[-m_0 c^2\{1-(\dot{x}^2+\dot{y}^2+\dot{z}^2)/c^2\}^{1/2}]$$

on the right hand side because it is identically zero. Equation (3.69) then becomes

$$\frac{d}{dt}\frac{\partial}{\partial \dot{x}}\left(-m_0c^2\{1-(\dot{x}^2+\dot{y}^2+\dot{z}^2)/c^2\}^{1/2} + \frac{e}{c}(\dot{x}A_x + \dot{y}A_y + \dot{z}A_z) - e\phi\right)$$

$$= \frac{\partial}{\partial x}[-e\phi + \frac{e}{c}\mathbf{V}\cdot\mathbf{A} - m_0 c^2\{1-(\dot{x}^2+\dot{y}^2+\dot{z}^2)/c^2\}^{1/2}].$$

or
$$\frac{d}{dt}[\frac{\partial}{\partial \dot{x}}(-m_0 c^2\sqrt{1-V^2/c^2} + \frac{e}{c}\mathbf{V}\cdot\mathbf{A} - e\phi)]$$

$$= \frac{\partial}{\partial x}(-m_0 c^2\sqrt{1-V^2/c^2} + \frac{e}{c}\mathbf{V}\cdot\mathbf{A} - e\phi). \tag{3.70a}$$

Similarly, for y- and z-components of equation (3.67), we have

$$\frac{d}{dt}[\frac{\partial}{\partial \dot{y}}(-m_0 c^2\sqrt{1-V^2/c^2} + \frac{e}{c}\mathbf{V}\cdot\mathbf{A} - e\phi)]$$

$$= \frac{\partial}{\partial y}(-m_0 c^2\sqrt{1-V^2/c^2} + \frac{e}{c}\mathbf{V}\cdot\mathbf{A} - e\phi) \tag{3.70b}$$

and
$$\frac{d}{dt}[\frac{\partial}{\partial \dot{z}}(-m_0 c^2\sqrt{1-V^2/c^2} + \frac{e}{c}\mathbf{V}\cdot\mathbf{A} - e\phi)]$$

$$= \frac{\partial}{\partial z}\left(-m_0 c^2\sqrt{1-V^2/c^2} + \frac{e}{c}\mathbf{V}\cdot\mathbf{A} - e\phi\right). \qquad (3.70c)$$

Comparing these equations with the Lagrange equations, viz.,

$$\frac{d}{dt}\frac{\partial L}{\partial \dot{q}_i} = \frac{\partial L}{\partial q_i},$$

we see that the Lagrangian L is given by

$$L = -m_0 c^2\sqrt{1-V^2/c^2} + \frac{e}{c}\mathbf{V}\cdot\mathbf{A} - e\phi.$$

In order that the relativistic Lagrangian for the system may reduce, in the limit $c\to\infty$, to conventional classical expression we may add the constant quantity m_0c^2 to the Lagrangian without affecting the Lagrange equations of motion so that

$$L = m_0c^2 - m_0 c^2\sqrt{1-V^2/c^2} + \frac{e}{c}\mathbf{V}\cdot\mathbf{A} - e\phi. \qquad (3.71)$$

This is the relativistic Lagrangian of a charged particle moving in an electromagnetic field.

The relativistic Hamiltonian of the particle can now be obtained from the equation

$$H(p_i, q_i) = \sum_i p_i \dot{q}_i - L, \quad \text{where} \quad p_i = \frac{\partial L}{\partial \dot{q}_i}.$$

Writing p_x, p_y, p_z for p_1, p_2, p_3 and x, y, z for q_1, q_2, q_3 respectively, we have

$$p_x = \frac{\partial L}{\partial \dot{x}} = \frac{\partial}{\partial \dot{x}}\left[m_0 c^2 - m_0 c^2\sqrt{1-V^2/c^2} + \frac{e}{c}\mathbf{V}\cdot\mathbf{A} - e\phi\right]$$

$$= \frac{\partial}{\partial \dot{x}}\left[-m_0 c^2\{1-(\dot{x}^2+\dot{y}^2+\dot{z}^2)/c^2\}^{1/2}\right] + \frac{e}{c}A_x$$

$$= m_0\dot{x}/\{1-(\dot{x}^2+\dot{y}^2+\dot{z}^2)/c^2\}^{1/2} + \frac{e}{c}A_x$$

$$= \frac{m_0}{\sqrt{1-V^2/c^2}}\dot{x} + \frac{e}{c}A_x.$$

We have similar expressions for p_y and p_z. Substituting the expressions for p_x, p_y, p_z and L in the defining relation for H, we get

$$H(\mathbf{p}, \mathbf{r}) = \sum_{x,y,z} p_x \dot{\mathbf{x}} - L$$

$$= \frac{m_0}{\sqrt{1 - V^2/c^2}} \dot{x}^2 + \frac{e}{c} \dot{x} A_x + \frac{m_0}{\sqrt{1 - V^2/c^2}} \dot{y}^2 + \frac{e}{c} \dot{y} A_y$$

$$+ \frac{m_0}{\sqrt{1 - V^2/c^2}} \dot{z}^2 + \frac{e}{c} \dot{z} A_z - L$$

$$= \frac{m_0}{\sqrt{1 - V^2/c^2}} V^2 + \frac{e}{c} \mathbf{V} \cdot \mathbf{A} - L$$

$$= \frac{m_0}{\sqrt{1 - V^2/c^2}} V^2 + \frac{e}{c} \mathbf{V} \cdot \mathbf{A} - m_0 c^2 + m_0 c^2 \sqrt{1 - V^2/c^2}$$

$$- \frac{e}{c} \mathbf{V} \cdot \mathbf{A} + e \phi$$

$$= \frac{m_0}{\sqrt{1 - V^2/c^2}} c^2 - m_0 c^2 + e \phi$$

or

$$H - e \phi + m_0 c^2 = \frac{m_0}{\sqrt{1 - V^2/c^2}} c^2$$

or

$$(H - e \phi + m_0 c^2)^2 = \frac{m_0^2}{1 - V^2/c^2} c^4. \tag{3.72}$$

We will now express the velocity V in terms of the generalized momenta. We have shown that

$$p_x = \frac{\partial L}{\partial \dot{x}} = \frac{m_0}{\sqrt{1 - V^2/c^2}} \dot{x} + \frac{e}{c} A_x .$$

This gives $p_x - \frac{e}{c} A_x = \frac{m_0}{\sqrt{1 - V^2/c^2}} \dot{x}.$

Similarly, $p_y - \frac{e}{c} A_y = \frac{m_0}{\sqrt{1 - V^2/c^2}} \dot{y}$

and $p_z - \frac{e}{c} A_z = \frac{m_0}{\sqrt{1 - V^2/c^2}} \dot{z}.$

Combining the last three equations, we get

$$\mathbf{p} - \frac{e}{c}\mathbf{A} = \frac{m_0}{\sqrt{1 - V^2/c^2}}\mathbf{V}.$$

Squaring both sides, we obtain

$$(\mathbf{p} - \frac{e}{c}\mathbf{A})^2 = \frac{m_0^2}{1 - V^2/c^2} V^2 = \frac{m_0^2}{1 - V^2/c^2} V^2 + m_0^2 c^2 - m_0^2 c^2$$

$$= \frac{m_0^2}{1 - V^2/c^2} c^2 - m_0^2 c^2$$

or
$$\frac{m_0^2}{1 - V^2/c^2} c^2 = (\mathbf{p} - \frac{e}{c}\mathbf{A})^2 + m_0^2 c^2. \tag{3.73}$$

Comparing equations (3.72) and (3.73), we get

$$(H - e\phi + m_0 c^2)^2 = c^2 (\mathbf{p} - \frac{e}{c}\mathbf{A})^2 + m_0^2 c^4$$

or
$$H = e\phi - m_0 c^2 + \sqrt{c^2 (\mathbf{p} - \frac{e}{c}\mathbf{A})^2 + m_0^2 c^4}. \tag{3.74}$$

This is the relativistic Hamiltonian of a charged particle moving in an electromagnetic field.

Variable Proper Mass

We have so far considered those cases only for which the proper mass of the given particle is constant. However, in certain cases, the proper mass of the particle may change. For instance, as a raindrop gathers moisture during its fall, its proper mass changes. Let us see how relativistic formulae can be generated in such cases.

Consider a particle of *variable proper mass* m_0. The 4-momentum of this particle is still defined by the equation $p_\mu = m_0 V_\mu$, but m_0 is no more a constant. The force f_μ is, therefore, given by

$$f_\mu = \frac{dp_\mu}{dt} = \frac{d}{dt}(m_0 V_\mu) = \frac{dm_0}{dt} V_\mu + m_0 \frac{dV_\mu}{dt}.$$

Multiplying both sides by V_μ and summing over μ, we obtain

$$V_\mu f_\mu = \frac{dm_0}{dt} V_\mu V_\mu + m_0 V_\mu \frac{dV_\mu}{dt}. \tag{3.75}$$

Now, from equation (3.3), we have

$$V_\mu V_\mu = -c^2.$$

Differentiating this with respect to t, we get

$$V_\mu \frac{dV_\mu}{dt} = 0.$$

Moreover, from equations (3.2e) and (3.19d), we have

$$V_\mu = \gamma(V)\,(\mathbf{V}, ic) \tag{3.2e$'$}$$

and

$$f_\mu = \left(\mathbf{f}, ic\frac{dm}{dt}\right). \tag{3.19d$'$}$$

Substituting these expressions for V_μ, f_μ, $V_\mu V_\mu$ and $V_\mu\, dV_\mu/dt$ in equation (3.75), we get

$$-c^2\frac{dm_0}{dt} = \gamma(V)\left(\mathbf{f}\cdot\mathbf{V} - c^2\frac{dm}{dt}\right) = \gamma(V)\left[\left(\mathbf{f}\cdot\mathbf{V} - \frac{d}{dt}(mc^2)\right]\right.$$

$$= \gamma(V)\left(\mathbf{f}\cdot\mathbf{V} - \frac{dE}{dt}\right)$$

or

$$\frac{dE}{dt} = \mathbf{f}\cdot\mathbf{V} + c^2\sqrt{1 - V^2/c^2}\,\frac{dm_0}{dt}. \tag{3.76}$$

Now dE/dt is the rate of increase of energy of the particle and $\mathbf{f}\cdot\mathbf{V}$ is the rate of work done on it by the applied force. Therefore $c^2\sqrt{1-V^2/c^2}\;dm_0/dt$ in equation (3.76) should be interpreted as the rate at which energy is taken from some external source.

Conservation of 4-Momentum for a System of Particles

So far we have been concerned with the dynamics of a single particle. Let us now consider the dynamics of a system of particles. We know that, in Newtonian mechanics, equation of motion for a system of particles is given by

$$\frac{d\mathbf{P}}{dt} = \mathbf{F}, \tag{3.77}$$

where **P** is the 3-momentum of the system and **F** is the force experienced by it. The force **F** can be split up into two parts, the external force acting on the system and the force of interaction between pairs of particles, so that we may write

$$\frac{d\mathbf{P}}{dt} = \mathbf{F}_{ext} + \mathbf{F}_{int}. \tag{3.78}$$

If no external force is acting on the system, then $\mathbf{F}_{ext} = 0$ and equation (3.78) reduces to

$$\frac{d\mathbf{P}}{dt} = \mathbf{F}_{int}. \tag{3.79}$$

But, according to Newton's third law of motion, action and reaction are equal and opposite, i.e., the forces which two particles exert on each other are equal and opposite, and act along the line joining them, so that for each pair of particles i and j, $\mathbf{F}_{ij} + \mathbf{F}_{ji}$ is zero. This means that $\mathbf{F}_{int}$, the sum of mutual forces between pairs of particles, vanishes:

$$\mathbf{F}_{int} = \sum_{\substack{i,j \\ i \neq j}} \mathbf{F}_{ij} = 0.$$

Then equation (3.79) reduces to $d\mathbf{P}/dt = 0$, so that the momentum of the system of particles is constant in time. That is, for an isolated system of particles, the law of conservation of momentum can be derived from Newton's second and third laws of motion.

Similarly, we can show that the law of conservation of energy for an isolated system of particles follows from Newton's laws.

This procedure, however, cannot be adopted in special relativity because, besides the fact that proper times of all the particles in the equation of motion of the system cause intricate problems and the third law of motion does not hold in general. To illustrate the last statement, consider two particles A and B separated by some distance and interacting with each other. When A exerts an action on B, then, owing to the finite speed of all signals, the reaction of B will not reach A instantaneously. Moreover, by the time this reaction reaches the position of A, this particle would have moved from that position. Therefore for any pair of particles i and j, the force of interaction $\mathbf{F}_{ij} + \mathbf{F}_{ji}$ will not be zero. Consequently, for a system of particles the sum total of forces of interaction between pairs of particles does not vanish; of course, even if $\mathbf{F}_{int}$ happens to be zero at a particular instant in one frame, due to relativity of simultaneity, it will not be so in other frames. However, if the interaction between pairs of particles takes place *only when they are in contact*, action and reaction will be equal and opposite. Third law of motion then holds good in all inertial frames because the simultaneity of coincident events is absolute.

We will now *assume* that the 4-momentum P_μ of an isolated system of particles is conserved. In fact, if the particles interact even when they are separated by some distance, then, unlike the non-relativistic situation, owing to the finite speed of the transmission of interaction, a fraction of the energy-momentum which has left some of the particles but has not yet reached other ones is in the space in between the particles. We have therefore to take into account this energy-momentum in the space-time region and write

$$P_\mu = P_\mu^{(\text{particles})} + P_\mu^{(\text{interaction})}.$$

That is, the 4-momentum P_μ of the system is equal to the sum of the 4-momenta of the particles and the 4-momenta in the interaction region. Since the measurements are made in a region where the interaction is negligible, for practical purposes $P_\mu = P_\mu^{(\text{particles})}$. This assumption is justified because its consequences are verified experimentally. Thus we may write

$$\sum_{i=1}^{N} p_\mu(i) = \sum_{j=1}^{N'} p_\mu'(j), \tag{3.80}$$

where N and N′ denote the number of the particles before and after the collision while $p_\mu(i)$ and $p_\mu'(j)$ are the 4-momenta of the ith and jth particles before and after the collision. Different letters have been used on the two sides of the equation because it is not necessary that in any interaction the number of interacting particles should remain the same. For instance, in the reaction $K^- + p \rightarrow p + \pi^- + K_0$, the number of particles before and after the interaction is different. Notice that equation (3.80), being a 4-vector equation, is covariant under Lorentz transformations and therefore is valid in *all* inertial frames. In this article i and j run from 1, • • •, N and 1, • • •, N′ respectively. Equation (3.81) may be written as

$$\Sigma\, [p_1(i), p_2(i), p_3(i), p_4(i)] = \Sigma\, [p_1'(j), p_2'(j), p_3'(j), p_4'(j)].$$

Equating the corresponding components on the two sides of this equation, we obtain

$$\Sigma\, p_1(i) = \Sigma\, p_1'(j) \tag{3.81a}$$

$$\Sigma\, p_2(i) = \Sigma\, p_2'(j) \tag{3.81b}$$

$$\Sigma\, p_3(i) = \Sigma\, p_3'(j) \tag{3.81c}$$

$$\Sigma\, p_4(i) = \Sigma\, p_4'(j). \tag{3.81d}$$

Since $p_1 = p_x$, $p_2 = p_y$, $p_3 = p_z$, etc., the first three equations may be put together in vector form:

$$\Sigma\, \mathbf{p}(i) = \Sigma\, \mathbf{p}'(j), \tag{3.82}$$

where $\mathbf{p} = (p_1, p_2, p_3) = (p_x, p_y, p_z)$

and $\mathbf{p}' = (p_1', p_2', p_3') = (p_x', p_y', p_z')$.

This is the **law of conservation of 3-momentum**; it shows that the 3-momentum of an isolated system of particles remains unchanged on collision.

Since $p_4 = imc$, equation (3.81d) yields

$$\Sigma\, m(i) = \Sigma m'(j). \tag{3.83}$$

This equation shows that the *relativistic mass of a system is always conserved.* Multiplying above equation by c^2 and using the relation $E = m\,c^2$, we get

$$\Sigma\, E(i) = \Sigma\, E'(j). \tag{3.84}$$

This equation expresses the **law of conservation of energy**. We may write the above equation as

$$\Sigma\; [m_0(i)\, c^2 + T(i)] = \Sigma\, [m_0(j)\, c^2 + T'(j)]. \tag{3.85}$$

This shows that the rest mass and the kinetic energy of a system need not be separately conserved. It is the total energy of the system which is always conserved. This is known as the **law of conservation of mass and energy**, or **mass-energy conservation law.**

The last three equations actually express the same conservation law.

Collision and Decay of Particles

Collisions and decay processes have played a significant role in the development of nuclear and elementary particle physics. At high energy, these processes can be investigated by making use of the laws of conservation of energy and momentum, and the relativistic expressions for energy and other physical quantities. We shall now discuss such processes in detail.

Collisions are usually observed in a frame of reference in which the target is at rest. This frame of reference is called the laboratory frame and is abbreviated as LAB frame. However, the algebra of collision problems is very much simplified if the calculations are made in a frame of reference in which the 3-momentum of the system is zero, even though the individual particles may be moving in this frame. We will first show that such a frame of reference does exist. To prove this, consider a system of free particles. Let **P** denote the initial 3-momentum of the system in the laboratory frame, say, S. We can take, without any loss of generality, the x-axis of the laboratory frame S along **P**. Let us consider a frame of reference S' having its x'-axis coincident with x-axis of the laboratory frame S and moving along the positive direction of the x-axis with a uniform velocity relative to S. If the relative velocity of the two frames is sufficiently large, then by virtue of the relativistic law for the addition of velocities, the x'-component of velocity and hence of momentum of every particle will have a negative value, so that the x'-component of the 3-momentum of the system is also negative. Moreover, as **P**, the 3-momentum of the system in S, is along x-axis, its y- and z-components will be zero. Since y- and z-components of the momentum of a particle do not change under a Lorentz transformation, the y'- and z'-components of the momentum of the system will also be zero. Hence the momentum of the system in the S' frame will be negative. Between these *two* extremes, there will exist a frame of reference for which the total momentum of the system is zero. This frame of reference is called the centre of momentum frame of the system or, loosely, the centre of mass frame. It is abbreviated as CM frame.

Let us now determine the velocity **v** of the CM frame S^* relative to the laboratory frame S, it being assumed that the two frames are in standard configuration. If E is the energy and p_1, p_2, p_3 are the components of the 3-momentum **p** of one of the particles of the system in the laboratory frame, and letters with asterisks indicate the corresponding quantities in the CM frame S^*, then the components of the 4-momentum of this particle in the two frames are related by

$$p_1^* = \gamma\,(p_1 + i\frac{v}{c}\,p_4) = \gamma\,(p_1 - \frac{v}{c^2}E) \tag{3.86a}$$

$$p_2^* = p_2 \tag{3.86b}$$

$$p_3^* = p_3 \tag{3.86c}$$

$$p_4^* = \gamma\,(p_4 - i\frac{v}{c}\,p_1). \tag{3.86d}$$

The last equation may be put in the form

$$E^* = \gamma\,(E - v\,p_1). \tag{3.86e}$$

Therefore the transformation law for the components of 3-momentum and the energy of a system of particles is given by

$$P_1^* = \Sigma\,p_1^* = \Sigma\,\gamma\,(\,p_1 - \frac{v}{c^2}E\,) = \gamma\,(\,\Sigma\,p_1 - \frac{v}{c^2}\Sigma\,E\,)$$

$$= \gamma\,(\,P_1 - \frac{v}{c^2}\varepsilon\,) \tag{3.87a}$$

$$P_2^* = \Sigma\,p_2^* = \Sigma\,p_2 = P_2 \tag{3.87b}$$

$$P_3^* = \Sigma\,p_3^* = \Sigma\,p_3 = P_3 \tag{3.87c}$$

$$\varepsilon^* = \Sigma\,E^* = \Sigma\,\gamma\,(E - v\,p_1) = \gamma\,(\Sigma\,E - v\,\Sigma p_1) = \gamma\,(\varepsilon - v\,P_1), \tag{3.87d}$$

where ε and ε^* denote the energies of the system in the LAB and CM frames, respectively. Since S* is the CM frame, all components of the 3-momentum of the system in this frame must be zero so that we have

$$P_1^* = P_2^* = P_3^* = 0.$$

Equations (3.87a, b, c) then give

$$P_1 = \frac{v}{c^2}\varepsilon,$$

$$P_2 = 0,$$

$$P_3 = 0.$$

These equations show that the velocity $\mathbf{v}$ = (v, 0, 0) of the CM frame relative to the laboratory frame is given by $\mathbf{v} = c^2\mathbf{P}/\mathcal{E}$, where $\mathbf{P} = \Sigma\, \mathbf{p}$. Equations (3.87) show that the quantity ($\mathbf{P}$, i $\mathcal{E}/c$) transforms as a 4-vector for the whole system in going from S to S*.

Since ($\mathbf{P}$, i $\mathcal{E}/c$) is a 4-vector, the square of its magnitude should be invariant. Therefore, we have

$$P^2 - \frac{\mathcal{E}^2}{c^2} = P^{*2} - \frac{\mathcal{E}^{*2}}{c^2} .$$

Thus this relation is valid even for a system of particles. For the CM frame, for which $\mathbf{P}^* = 0$, this relation reduces to

$$P^2 - \frac{\mathcal{E}^2}{c^2} = -\frac{\mathcal{E}^{*2}}{c^2} .$$

The quantity $\mathcal{E}^*/c^2$ is called the **rest mass of the system of non-interacting particles** in the CM frame because in this frame the 3-momentum $\mathbf{P}^*$ of the system is zero. It is denoted by M*. Now, we have

$$\mathcal{E}^* = E^*(1) + E^*(2) + \cdots + E^*(n).$$

This gives

$$M^* c^2 = M_1^* c^2 + M_2^* c^2 + \cdots + M_n^* c^2$$

or $$M^* = M_1^* + M_2^* + \cdots + M_n^*.$$

That is, the rest mass M* of the system of particles in the CM frame is equal to the sum of the relativistic masses of the particles in the CM frame. Since the square of the magnitude of a 4-vector is an invariant quantity, the rest mass M*, being equal to the negative of $P_\mu^* P_\mu^*$, is invariant with respect to a Lorentz transformation.

The **relativistic mass of the system** in an arbitrary frame is defined by the expression $\Sigma E/c^2$ and is denoted by M.

Elastic Collision of Identical Particles

As an illustration of the application of the laws of conservation of energy and momentum in relativistic collisions, we shall solve the problem of

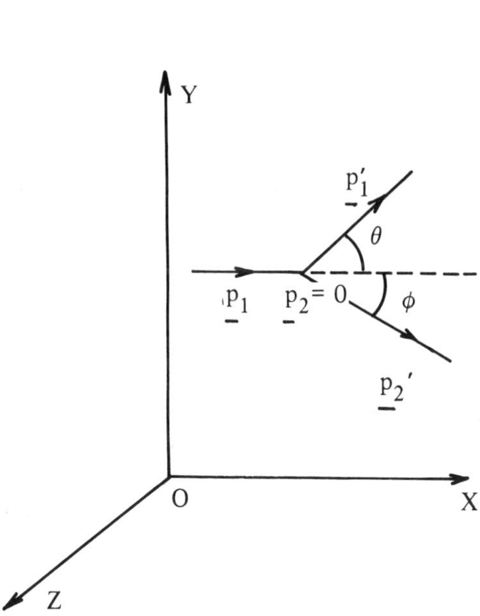

Fig. 3.1a. Scattering of identical particles in the LAB frame

the elastic collision of identical particles, say the collision of fast moving electrons with electrons at rest in the laboratory frame of reference. The main point of interest in this problem is the *calculation of the angle between the paths of motion of the electrons after the collision*, as this angle can be measured experimentally. To compute this angle, we proceed as follows. Let m_0 be the rest mass of each of the two electrons. Since the collision has been assumed to be elastic, the rest masses of the particles are not changed on collision. Let $\mathbf{p}(1)$, $\mathbf{p}(2) = 0$, E_1, E_2 and $\mathbf{p}'(1)$, $\mathbf{p}'(2)$, E_1', E_2' be the momenta and energies of the two electrons in the laboratory frame before and after the collision. Let us choose the laboratory frame S such that its x-axis is parallel to $\mathbf{p}(1)$ and that the momentum vector $\mathbf{p}'(1)$ lies in its xy-plane. Since linear momentum is always conserved, the momentum vector $\mathbf{p}'(2)$ should also be in the xy-plane. After collision, the two particles leave in directions making angles θ and ϕ with the initial direction of motion of the incident particle. This is shown in Fig.3.1a. The direct computation of angles θ and ϕ in the laboratory frame by using the laws of conservation of energy and momentum is very complicated. We shall therefore solve this problem in the CM frame in which the calculations become simple and then using an appropriate Lorentz transformation obtain the result in the laboratory frame. Notice that the variables in the CM frame would be denoted by letters with asterisks.

We shall first derive certain results in the CM frame which would be used in calculations later on.

By definition, the 3-momentum of the system in the CM frame is zero, so that

$$\mathbf{p}^*(1) + \mathbf{p}^*(2) = 0$$

or

$$\frac{m_0}{\sqrt{1 - V^{*2}(1)/c^2}} \mathbf{V}^*(1) = -\frac{m_0}{\sqrt{1 - V^{*2}(2)/c^2}} \mathbf{V}^*(2),$$

where $\mathbf{V}^*(1)$ and $\mathbf{V}^*(2)$ are the velocities of the colliding particles. Squaring and simplifying, we get

$$V^*(1) = V^*(2) = V^*, \text{ say.} \tag{3.88}$$

Thus, before collision, the particles in the CM frame must move with the same speed V^*. Of course, as the sum of their momenta is zero, they must be moving in opposite directions. Moreover, as the two particles of equal rest masses are moving with the same speed V^*, their relativistic masses and hence their energies must be equal:

$$E_1^* = E_2^* = E^*, \text{ say} \tag{3.89}$$

By virtue of the law of conservation of momentum, the 3-momentum of the system in the CM frame would remain zero even after collision. Hence, as before, particles would rebound with equal and opposite velocities and their energies would be equal. We denote the values of speed and energy of each of the particles emerging on collision by V'^* and E'^* respectively.

Now from the law of conservation of energy in the CM frame, we must have

$$E_1^* + E_2^* = E_1'^* + E_2'^*. \tag{3.90}$$

Substituting the values of E_1^*, E_2^*, $E_1'^*$ and $E_2'^*$ from equations (3.88) and (3.89) in equation (3.90), we get

$$E'^* = E^*. \tag{3.91}$$

That is, in the CM frame each one of the identical particles has the same energy before as well as after the collision. As the rest masses are not changed in an elastic collision, we have

$$\frac{m_0}{\sqrt{1 - V'^{*2}/c^2}}\, c^2 = \frac{m_0}{\sqrt{1 - V^{*2}/c^2}}\, c^2 .$$

or

$$V'^* = V^*. \tag{3.92}$$

That is, in the CM frame, a collision between two electrons does not change their speed; it merely causes a change in the direction of motion in the CM frame of the electrons. We conclude that, in the CM frame, two *identical* particles move with the *same speed* before as well as after an elastic collision between them. This is illustrated in Fig.3.1b.

Let $p_\mu^{*\prime}$ and p_μ' be the 4-momenta of any one of the final particles in the CM and LAB frames respectively. Then the inverse transformation law for 4-momentum gives

$$\begin{aligned}
p_1' &= \gamma(V^*)\,(p_1^{*\prime} - i\frac{V^*}{c}\, p_4^{*\prime}) \\
p_2' &= p_2^{*\prime} \\
p_3' &= p_3^{*\prime} \\
p_4' &= \gamma(V^*)\,(p_4^{*\prime} + i\frac{V^*}{c}\, p_1^{*\prime}).
\end{aligned} \tag{3.93}$$

We are now well equipped to determine the expressions for the angles θ and ϕ in the laboratory frame.

From Fig.3.1, we have

$$\tan\theta = \frac{p_y'(1)}{p_x'(1)} = \frac{p_2'(1)}{p_1'(1)} .$$

By using the transformation equations (3.93), we have

$$\tan\theta = \frac{p_2^{*\prime}(1)}{\gamma(V^*)\,(p_1^{*\prime}(1) - i\frac{V^*}{c}\, p_4^{*\prime}(1))}$$

$$= \frac{p_2^{*\prime}(1)}{\gamma(V^*)\,(p_1^{*\prime}(1) + \frac{V^*}{c^2}\,E_1^{*\prime})}\,. \tag{3.94a}$$

Similarly, we have

$$\tan\phi = -\frac{p_y'(2)}{p_x'(2)} = -\frac{p_2'(2)}{p_1'(2)}$$

$$= -\frac{p_2^{*\prime}(2)}{\gamma(V^*)\,(p_1^{*\prime}(2) + \frac{V^*}{c^2}\,E_1^{*\prime})}$$

$$= \frac{p_2^{*\prime}(1)}{\gamma(V^*)\,(-p_1^{*\prime}(1) + \frac{V^*}{c^2}\,E_1^{*\prime})}\,, \tag{3.94b}$$

where we have made use of the fact that in the CM frame the energies of the two particles are equal and the corresponding components of their 3-momenta are equal and opposite.

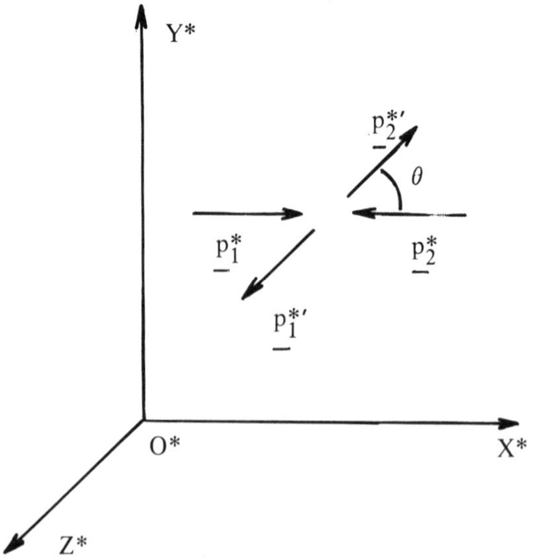

Fig. 3.1b. Scattering of identical particles in the CM frame

Multiplying equations (3.94a) and (3.94b), we get

$$\tan\theta\tan\phi = \frac{p_2^{*\prime 2}(1)}{\gamma^2(V^*)\left[-p_1^{*\prime 2}(1) + \frac{V^{*2}}{c^4}E_1^{*\prime 2}\right]} \tag{3.95}$$

Problem

Show that

$$p_2^{*\prime 2}(1) = -p_1^{\prime *2}(1) + \frac{V^{*\prime 2}}{c^4}E^{*}{}_1{}^{\prime}. \tag{3.96}$$

By virtue of relation (3.96), equation (3.95) reduces to

$$\tan\theta\tan\phi = \frac{1}{\gamma^2(V^*)} = 1 - V^{*2}/c^2. \tag{3.97}$$

Now the CM is moving along the common x-axis with a speed V* relative to the LAB frame. Therefore the speed V of the incident particle moving along the x-axis in the LAB frame is related to its speed in the CM frame by the equation

$$V = \frac{V^* + V^*}{1 + \frac{V^* \cdot V^*}{c^2}} = \frac{2V^*}{1 + V^{*2}/c^2}. \tag{3.98}$$

Problem

Find the expression for $\gamma^2(V^*)$ in terms of V.

By making use of this expression, equation (3.97) can be written as

$$\tan\theta\tan\phi = \frac{2}{1 + \gamma(V)}. \tag{3.99}$$

Equation (3.99) shows that the product $\tan\theta\tan\phi$ is independent of the rest mass of the identical particles and is a function of the velocity of the incident particle alone.

According to relativistic mechanics, for all particles with non-zero rest mass, we have

$$\gamma(V) = \frac{1}{\sqrt{1 - V^2/c^2}} > 1.$$

Therefore, equation (3.99) yields

$$\tan\theta \tan\phi < 1.$$

This means that $\theta + \phi$ must be less than $\pi/2$. This has been verified by Champion[3] by taking cloud chamber pictures of the collisions of fast moving electrons with electrons which are at rest in the chamber. It turns out that in many cases $\theta + \phi$ is considerably smaller than $\pi/2$, generally by about ten degrees or more. In fact, the observed values for $\theta + \phi$ were in very good agreement with the predictions of equation (3.99).

The corresponding classical relation is found by taking the limit $V/c \to 0$. Equation (3.99) then gives

$$\tan\theta \tan\phi = 1,$$

so that

$$\tan(\theta + \phi) = \frac{\tan\theta + \tan\phi}{1 - \tan\theta \tan\phi} \to \infty$$

or

$$\theta + \phi = \frac{1}{2}\pi. \tag{3.100}$$

This is the well-known result of Newtonian mechanics: after elastic collision, in the laboratory frame, the directions of motion of the identical particles are at right angles to each other. This is found to be true, for example, in the case of low-energy α-particle collisions in the laboratory where the velocities of the particles are very small compared to c.

The product $\tan\theta \tan\phi$ can also be expressed in terms of energy. If E is the energy of the incident particle in the LAB frame, then

$$E = \frac{m_0}{\sqrt{1 - V^2/c^2}} c^2 = m_0 \gamma(V) c^2.$$

Therefore, $\gamma(V) = E/m_0c^2$. Substituting this expression for $\gamma(V)$ in equation (3.99), we get

$$\tan\theta \tan\phi = \frac{2}{1 + \dfrac{E}{m_0c^2}} = \frac{2 m_0 c^2}{E + m_0 c^2}. \tag{3.101}$$

Threshold Energy Required for a Reaction, and Energy Available for the Creation of Particles which may Emerge on Collision

Beams of fast moving particles are produced by accelerators. These fast moving particles possess high energy by virtue of their motion and impinge on targets which are usually taken as stationary in the laboratory. Such collisions result in the production of particles which may or may not be identical with the colliding particles. These reactions play a significant role in the study of the properties of elementary particles. Sometimes a physicist is interested in a particular reaction and wants to know the conditions under which this reaction will actually occur. For that he is concerned with two problems. Firstly, what is the minimum energy required by the colliding particles for a particular reaction to be possible? For, it would be futile to try to observe a particular reaction, say

$$a + b \rightarrow c + d + e,$$

unless the colliding particles possess sufficient energy for the reaction to be possible. Secondly, is that much energy really available? We answer these questions one by one.

Consider any reaction. If we examine it from the LAB frame, then, unless the initial momentum of the system is zero, all the particles emerging after collision cannot be at rest. This is so because if the initial momentum of the system is different from zero and all the particles formed on collision are at rest, the final momentum of the system would be zero and this would violate the law of conservation of momentum. However, since the momentum of a system of particles in the CM frame is always zero, when examined from this frame all the particles emerging after collision can be at rest. Therefore, in the CM frame the minimum energy required for the creation of emerging particles is equal to the sum of the rest mass energies of these particles. This energy is called the **threshold energy** for the reaction. By making use of the invariance of the magnitude of 4-momentum, the minimum energy required for the creation of these particles in the laboratory can be calculated.

Next we have to examine whether this much energy is available or not for conversion into the rest masses of the emerging particles. Again we first examine this problem from the LAB frame. Unless the initial momentum of the system is zero, all the initial energy of the system is not available for conversion into the rest masses of the particles emerging after collision. This is so because if the initial momentum is different from zero and all the initial energy is converted into the rest masses of the particles formed by collision, all the emerging particles would be at rest as no energy would be available for their motion and therefore the final momentum of the system would be zero. This would violate the law of conservation of momentum. However, if we examine the reaction from the CM frame, the initial and consequently the final momenta of the system are always zero. This may occur when all the particles emerging on collision are stationary. Hence the total energy of the system before the collision can be converted into the rest masses of the particles emerging after the collision. In other words, all the CM energy of a system of particles before collision is available for the creation of the rest masses of the particles emerging after collision. If we know the initial energy and momentum of the system in the laboratory, then by making use of the invariance of the magnitude of the momentum 4-vector, we can calculate the initial energy of the system in the CM frame. It is this energy which is available for the rest masses of the particles which may emerge on collision.

Thus in the CM frame the minimum energy required for a particular reaction to take place is equal to the sum of the rest mass energies of the particles emerging after collision. The corresponding minimum value of energy in the LAB frame can be computed by making use of the energy momentum transformation law. The energy which is available for the rest masses of the emerging particles is equal to the total energy of the colliding particles in the CM frame. If we know the energies and momenta of the colliding particles in the laboratory frame, we can calculate the corresponding energy in the CM frame; this is the energy available for the rest masses of the particles which emerge on collision.

To find an expression for the minimum energy necessary for a particular reaction to be possible in the laboratory, consider two particles "a" and "b" of rest masses m_a and m_b, energies E_a and E_b and 3-momenta $\mathbf{p}(a)$ and $\mathbf{p}(b) = 0$ in the laboratory frame, colliding with each other to create particles, which may or may not be identical with the colliding particles, according to the reaction

$$a + b \rightarrow c + d + \cdots$$

As mentioned earlier, letters with asterisks would give the corresponding energies and momenta in the CM frame. Now, in the CM frame, the minimum energy required for this reaction is

$$\mathcal{E}^* = m_c c^2 + m_d c^2 + \cdots, \tag{3.102}$$

where m_c , m_d, • • • are the rest masses of the particles emerging on collision. If P_μ is the 4-momentum, **P** the 3-momentum and $\mathcal{E}$ the energy of the system of two particles in the laboratory frame corresponding to the threshold energy in the CM frame, then by virtue of the invariance of the magnitude of a 4-vector, we have

$$P^2 - \frac{\mathcal{E}^2}{c^2} = -\frac{\mathcal{E}^{*2}}{c^2}$$

or

$$\mathcal{E}^{*2} = \mathcal{E}^2 - c^2 P^2 \tag{3.103}$$

By substituting the expression for $\mathcal{E}^*$ from equation (3.102) in equation (3.103), we get

$$\mathcal{E}^2 = (m_c + m_d + \cdots)^2 c^4 + c^2 \mathbf{p}^2(a), \tag{3.104}$$

because the particle "b" is at rest in the laboratory. But, in view of equation (3.45), we have

$$c^2 \mathbf{p}^2(a) + m_a^2 c^4 = E_a^2 = (\mathcal{E} - E_b)^2 = (\mathcal{E} - m_b c^2)^2$$

$$= \mathcal{E}^2 + m_b^2 c^4 - 2m_b c^2 \mathcal{E}. \tag{3.105}$$

Substituting the expression for $c^2 \mathbf{p}^2(a)$ from equation (3.105) in equation (3.104), we get

$$\mathcal{E}^2 = (m_c + m_d + \cdots)^2 c^4 + \mathcal{E}^2 + m_b^2 c^4 - 2 m_b c^2 \mathcal{E} - m_a^2 c^4$$

or

$$\mathcal{E} = [(m_c + m_d + \cdots)^2 - m_a^2 + m_b^2] \frac{c^2}{2m_b}. \tag{3.106}$$

This is the *minimum energy required in the laboratory* for the reaction

$$a + b \rightarrow c + d + \cdots$$

to be possible. This analysis shows that a particular reaction

$$a + b \rightarrow c + d + \cdots$$

is possible only if the available energy is either equal to or greater than the minimum energy $(m_c + m_d + \cdot \cdot \cdot) c^2$.

Problem

Show that

$$E_a = \frac{\varepsilon^{*2} - m_a^2 c^4 - m_b^2 c^4}{2 m_b c^2} \tag{3.107}$$

and

$$\varepsilon = \frac{\varepsilon^{*2} - m_a^2 c^4 - m_b^2 c^4}{2 m_b c^2} . \tag{3.108}$$

If the final system consists of two particles "c" and "d" only, i.e., we are considering the reaction

$$a + b \rightarrow c + d,$$

then equation (3.106) reduces to

$$\varepsilon = \frac{[(m_c + m_d)^2 - m_a^2 + m_b^2] c^2}{2 m_b} . \tag{3.109}$$

We note from equations (3.106) and (3.107) that the corresponding kinetic energy T_a of the incident particle "a" is given by

$$T_a = E_a - m_a c^2$$

$$= \frac{[(m_c + m_d)^2 - m_a^2 - m_b^2 - 2 m_a m_b]c^4}{2 m_b c^2}$$

$$= \frac{[(m_c + m_d)^2 - (m_a + m_b)^2] c^2}{2 m_b} . \tag{3.110}$$

This shows that if the target is at rest, then, in order that the reaction $a + b \rightarrow c + d$ may occur, the minimum kinetic energy possessed by the incident particle should be given by equation (3.110). If the kinetic energy of the incident particle exceeds the threshold kinetic energy, the excess energy is shared by the particles emerging after the collision as their kinetic energy.

Let us next find the energy which is available for producing the rest masses of particles which may emerge on the collision of particles "a" and "b". Let $\mathcal{E}$ be the energy and **P** the momentum of the system in the laboratory frame and $\mathcal{E}^*$, the energy of the system in the CM frame. Then, as we have already discussed , all the energy $\mathcal{E}^*$ given by

$$\mathcal{E}^{*2} = \mathcal{E}^2 - c^2 \mathbf{P}^2 \qquad (3,103')$$

is available for producing the rest masses of the emerging particles. However, it is convenient for actual calculations to express the energy $\mathcal{E}$ and the momentum **P** of the system in terms of the energies E_a and E_b and momenta **p**(a) and **p**(b) of the individual particles in the laboratory frame, so that equation (3.103′) may be written as

$$\begin{aligned}\mathcal{E}^{*2} &= (E_a + E_b)^2 - c^2 [\mathbf{p}(a) + \mathbf{p}(b)]^2 \\ &= E_a^2 + E_b^2 + 2 E_a E_b - c^2 [\mathbf{p}^2(a) + \mathbf{p}^2(b) + 2 \mathbf{p}(a) \cdot \mathbf{p}(b)]. \qquad (3.111)\end{aligned}$$

But, in view of equation (3.45), we have

$$E_a^2 = c^2 \mathbf{p}^2(a) + m_a^2 c^4 \quad \text{and} \quad E_b^2 = c^2 \mathbf{p}^2(b) + m_b^2 c^4.$$

Substituting these expressions for $E_a^2 - c^2 \mathbf{p}^2(a)$ and $E_b^2 - c^2 \mathbf{p}^2(b)$ in equation (3.111), we get

$$\mathcal{E}^{*2} = (m_a^2 + m_b^2) c^4 + 2 [E_a E_b - c^2 \mathbf{p}(a) \cdot \mathbf{p}(b)]. \qquad (3.112)$$

If one of the particles, say "b", is at rest in the laboratory, then $\mathbf{p}(b) = 0$ and $E_b = m_b c^2$ and the above equation reduces to

$$\mathcal{E}^{*2} = (m_a^2 + m_b^2) c^4 + 2 m_b c^2 E_a. \qquad (3.113)$$

Problem

Show that equation (3.113) may be put in the form

$$\mathcal{E}^{*2} = (m_a^2 - m_b^2) c^4 + 2 m_b c^2 \mathcal{E}. \qquad (3.114)$$

Equation (3.113) shows that knowing the values of m_a, m_b and E_a, the available energy ε^* in the CM frame can be calculated. For incident particles of very high energy, $m_a^2c^4$ and $m_b^2c^4$ can be neglected as compared to E_a so that equation (3.113) reduces to

$$\varepsilon^{*2} = 2\, m_b\, c^2\, E_a$$

or $$\varepsilon^* = \sqrt{2\, m_b}\; c\sqrt{E_a}.$$

This equation shows that if the energy of the incident particle a in the laboratory is increased, then, ε^*, the amount of energy that is available in the CM frame for conversion into the rest masses of particles emerging on collision increases as the square root of the incident energy in the laboratory.

Problem

Using equation (3.102), show that

$$T^{*2} + 2\,(m_a + m_b)\, c^2\, T^* = 2\, m_2\, c^2\, T \tag{3.115}$$

where T and T* are the kinetic energies of the system in the LAB and CM frames, respectively.

Problem

Antiprotons $\bar{p}$ may be produced by bombarding hydrogen with protons p according to the reaction

$$p + p \rightarrow p + p + (p + \bar{p}).$$

With what minimum kinetic energy must the bombarding proton strike the proton at rest in order for this reaction to occur?

Solution

In the CM frame, the minimum energy required for the creation of particles emerging after proton-proton collision is equal to the sum of their rest mass energies: $\varepsilon^* = 4\, m_p\, c^2$, where m_p is the mass of a proton or an antiproton. Therefore, in the laboratory frame, the minimum energy required for the creation of pp and $p\bar{p}$ pairs is, according to equation (3.106), given by

$$\mathcal{E} = \frac{[(4\ m_p)^2 - m_p^2 + m_p^2]\ c^2}{2\ m_p} = 8\ m_p\ c^2.$$

Hence the minimum LAB energy which the incident proton may possess so that the above reaction may occur is

$$E_a = \mathcal{E} - E_b = 8\ m_p\ c^2 - m_p\ c^2 = 7\ m_p\ c^2. \tag{3.116}$$

The corresponding kinetic energy T of the incident protons is given by

$$T_a = E_a - m_p\ c^2 = 6\ m_p\ c^2 = 6 \times 938\ \text{MeV} = 5.6\ \text{GeV} \tag{3.117}$$

Problem

Discuss whether the motion of the nucleons in the target enhances or reduces the threshold energy in the laboratory frame.

Problem

Calculate the available energy when a proton beam collides with protons at rest.

Solution

Since both the incident particle and the target are protons, we have $m_a = m_b = m_p$, so that equation (3.105) yields the following result for the available energy $\mathcal{E}^*$:

$$\mathcal{E}^{*2} = 2\ m_p^2\ c^4 + 2\ m_p\ c^2\ E_a. \tag{3.118}$$

If the kinetic energy of the incident particles is much greater than its rest mass energy, we may ignore $2\ m_p^2\ c^4$ as compared with $2\ m_p\ c^2\ E_a$ and write the above equation as

$$\mathcal{E}^* = \sqrt{2\ m_p}\ c \sqrt{E_a} \tag{3.119}$$

Thus for $E_a = 25$ GeV, $\mathcal{E}^* = 6.85$ GeV. That is if the stationary protons in the laboratory are bombarded with high energy protons, only a fraction of the energy is available for the creation of the rest masses of new particles.

Equation (3.119) shows that if we increase the total energy of the incident protons from 25 GeV to 100 GeV (and this would cause a tremendous increase in the cost of construction of an accelerator), the energy available for the rest masses of new particles would only be approximately doubled. To overcome this difficulty, an accelerator, the intersecting storage rings (ISR), has been built in which the beams of particles with equal momenta and coming from opposite directions collide with each other so that the laboratory frame also serves as the CM frame and all the energy of the colliding particles is available for the creation of the rest masses of the emerging particles.

Decay Processes in High Energy Physics

We shall now consider the dynamics of the decay of a particle "d" into two particles "a" and "b":

$$d \rightarrow a + b$$

Let m_d, m_a and m_b be the rest masses, E_d, E_a and E_b the energies, and $\mathbf{p}(d)$, $\mathbf{p}(a)$ and $\mathbf{p}(b)$ the 3-momenta of these particles. Then according to the laws of conservation of energy and momentum, we must have

$$E_d = E_a + E_b \tag{3.120}$$

$$\mathbf{p}(d) = \mathbf{p}(a) + \mathbf{p}(b). \tag{3.121}$$

Suppose first that at the time of its decay, the particle "d" is at rest. Then $\mathbf{p}(d) = 0$ and equation (3.121) reduces to

$$\mathbf{p}(a) = -\,\mathbf{p}(b) \tag{3.122}$$

i.e., the momenta of the products of decay are equal and opposite. Squaring equation (3.122), we obtain

$$[\mathbf{p}(a)]^2 = [\mathbf{p}(b)]^2 = \flat^2, \text{ say.} \tag{3.123}$$

However, by using equation (3.45), viz., $E^2 = m_0^2 c^4 + \mathbf{p}^2 c^2$, we obtain

$$E_d^2 = m_d^2 c^4 \qquad \text{or} \qquad E_d = m_d c^2 \tag{3.124a}$$

$$E_a^2 = m_a^2 c^4 + [p(a)]^2 c^2 = m_a^2 c^4 + p^2 c^2 \qquad (3.124b)$$

$$E_b^2 = m_b^2 c^4 + [p(b)]^2 c^2 = m_b^2 c^4 + p^2 c^2. \qquad (3.124c)$$

Subtracting equation (3.124c) from equation (3.124b), we get

$$E_a^2 - E_b^2 = (m_a^2 - m_b^2)\, c^4. \qquad (3.125)$$

Eliminating E_d between equations (3.120) and (3.124a), we have

$$E_a + E_b = m_d\, c^2. \qquad (3.126)$$

Substituting for $E_a + E_b$ in equation (3.125) from equation (3.126), we obtain

$$E_a - E_b = \frac{m_a^2 - m_b^2}{m_d}\, c^2. \qquad (3.127)$$

Adding equations (3.126) and (3.127), we get

$$E_a = \frac{m_d^2 + m_a^2 - m_b^2}{2\, m_d}\, c^2. \qquad (3.128)$$

This gives the total energy of the decay product "a". The total energy of the decay product "b" is obtained by subtracting equation (3.127) from equation (3.126):

$$E_b = \frac{m_d^2 + m_b^2 - m_a^2}{2m_d}\, c^2. \qquad (3.129)$$

Suppose next that the particle "d" *decays in flight* such that the emerging particles "a" and "b" move in directions making an angle ϕ with each other. Then squaring equation (3.121), we get

$$p^2(d) = p^2(a) + p^2(b) + 2\, p(a)\, p(b) \cos\phi, \qquad (3.130)$$

where $p(i)$ is the magnitude of the 3-momentum of the particle "i". Again by using equation (3.45), viz., $E^2 = m_0^2 c^4 + p^2 c^2$, we may write

$$E_d^2 = m_d^2 c^4 + p^2(d)\, c^2 \qquad (3.131a)$$

$$E_a^2 = m_a^2 c^4 + \mathbf{p}^2(a) c^2 \tag{3.131b}$$

$$E_b^2 = m_b^2 c^4 + \mathbf{p}^2(b) c^2 \tag{3.131c}$$

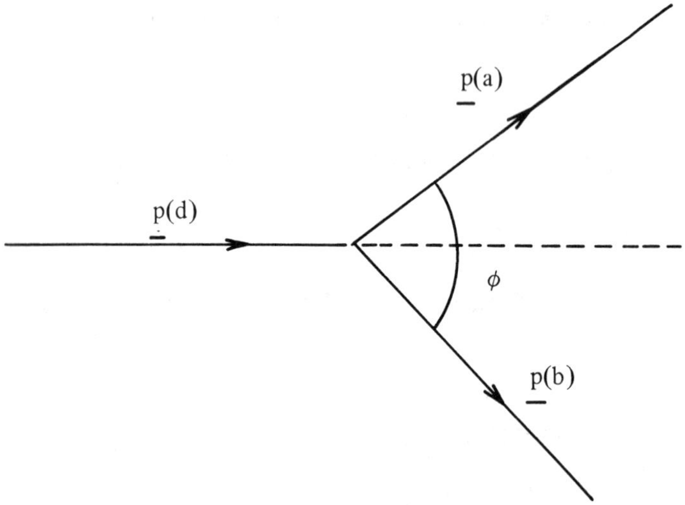

Fig. 3.2. Decay of a particle

Substituting the expressions for $\mathbf{p}^2(d)$, $\mathbf{p}^2(a)$ and $\mathbf{p}^2(b)$ from equations (3.131) in equation (3.130), we get

$$E_d^2 - m_d^2 c^4 = E_a^2 - m_a^2 c^4 + E_b^2 - m_b^2 c^4 + 2\, \mathfrak{p}(a)\, \mathfrak{p}(b) \cos\phi \tag{3.132}$$

From equation (3.120), we get

$$E_d^2 = E_a^2 + E_b^2 + 2\, E_a\, E_b.$$

Substituting this expression for E_d^2 in equation (3.132), and simplifying, we get

$$\cos\phi = \frac{(m_a^2 + m_b^2 - m_d^2)\, c^4 + 2\, E_a\, E_b}{2\, \mathfrak{p}(a)\, \mathfrak{p}(b)}\, c^2. \tag{3.133}$$

Knowing the various quantities on the right hand side, we can calculate ϕ.

Particles of Zero Rest Mass

We know that the energy E and the momentum $p = |\mathbf{p}|$ of a particle are related by the equation

$$E^2 = m_0^2 c^4 + p^2 c^2 \tag{3.134}$$

$$\frac{E}{p} = \frac{c^2}{v} \tag{3.135}$$

The second relation has been obtained by eliminating m between equations $E = mc^2$ and $p = mv$. If a particle is moving with a velocity c, then, for $v = c$, equation (3.135) gives

$$E = c\,p \tag{3.136}$$

so that equation (3.134) yields $m_0 = 0$. This shows that the rest mass of a particle moving with the velocity of light must be zero. Photons and neutrinos are such particles. It must, however, be stressed that such particles actually can never come to rest. In fact, if a particle is moving with a velocity c in one frame, it would be moving with the same speed c in all inertial frames. It will be either moving with this speed or will cease to exist. Since according to the quantum theory, the energy of a photon is given by

$$E = h\,\nu \tag{3.137a}$$

where ν is the frequency of light and h is Planck's constant, we may write

$$p = \frac{E}{c} = \frac{h\nu}{c} \tag{3.137b}$$

The formulae derived above are frequently used in high energy physics. We shall illustrate this by a few examples.

Problem

A Λ^0 hyperon of rest mass m_Λ decays into two massive particles, a proton p of rest mass m_p and a meson of rest mass m_π according to the relation

$$\Lambda^0 \to \pi^- + p.$$

Assuming that the Λ^0 hyperon is stationary at the time of its decay, find the expressions for the total and kinetic energies of the proton and the pion.

Solution

Equations (3.128) and (3.129) yield

$$E_p = \frac{m_\Lambda^2 + m_p^2 - m_\pi^2}{2\, m_\Lambda}\, c^2$$

and

$$E_\pi = \frac{m_\Lambda^2 + m_\pi^2 - m_p^2}{2\, m_\Lambda}\, c^2.$$

The corresponding kinetic energies T_p and T_π are given by

$$T_p = E_p - m_p\, c^2 = \frac{(m_\Lambda - m_p)^2 - m_\pi^2}{2\, m_\Lambda}\, c^2$$

$$T_\pi = E_\pi - m_\pi\, c^2 = \frac{(m_\Lambda - m_\pi)^2 - m_p^2}{2\, m_\Lambda}\, c^2 .$$

Similar results may be obtained for

$$\Lambda^0 \rightarrow \pi^0 + n$$

$$\Xi^- \rightarrow \pi^- + \Lambda^0$$

$$\Omega^- \rightarrow \pi^- + \Xi^0$$

$$K^+ \rightarrow \pi^+ + \pi^0.$$

and many other decay processes.

Problem

A π^+ meson of rest mass m_π decays into a muon of rest mass m_μ and a neutrino of zero rest mass:

$$\pi^+ \rightarrow \mu^+ + \nu_\mu.$$

Assuming that π^+ meson is stationary at the time of its decay, find the expressions for the total and kinetic energies of the muon and the corresponding neutrino.

Solution

By virtue of relations (3.128) and (3.129), in the rest frame of the decaying particle, viz., π^+ meson, we have

$$E_\mu = \frac{m_\pi^2 + m_\mu^2}{2\, m_\pi} c^2$$

$$E_\nu = \frac{m_\pi^2 + m_\nu^2}{2\, m_\pi} c^2$$

$$T_\mu = E_\mu - m_\mu c^2 = \frac{(m_\pi - m_\mu)^2}{2\, m_\pi} c^2$$

and $$T_\nu = E_\nu - m_\nu c^2 = E_\nu = \frac{(m_\pi - m_\nu)^2}{2\, m_\pi} c^2 .$$

It may be noticed that the rest mass of the neutrino has been taken as zero.

Problem

Suppose that a π^0 meson of rest mass m_π decays into two photons of different frequencies:

$$\pi^0 \rightarrow \gamma + \gamma .$$

Calculate the angle between these photons.

Solution

The decay of a π^0 meson of rest mass m_π and momentum **p** into two photons of different frequencies ν_1 and ν_2 is shown in Fig.3.3. The rest mass of the decaying particle is $m_d = m_\pi$ where the rest masses of the products, m_a and m_b, are zero. The energies of the photons produced in this process are $E_a = h\nu_1$ and $E_b = h\nu_2$. The corresponding magnitudes of 3-momenta are $p_a = h\nu_1/c$ and $p_b = h\nu_2/c$. Substituting these expressions in equation (3.132), we get

$$\cos\phi = 1 - \frac{m_\pi^2 c^4}{2\, h^2 \nu_1 \nu_2} .$$

This can also be written as

$$\sin\frac{1}{2}\phi = \frac{m_\pi c^2}{2h\sqrt{\nu_1\nu_2}}. \tag{3.138}$$

This equation may be used to compute the rest mass of the π^0 meson if ϕ and ν_1, ν_2 are determined experimentally.

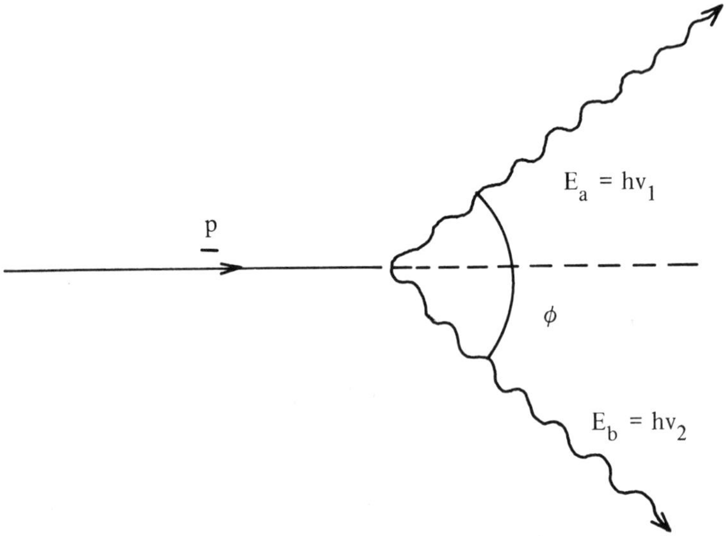

Fig. 3.3. Decay of π^0 into two photons

Compton Effect

When electromagnetic radiation of high frequency and consequently of high energy is incident on electrons of a light element in which the electrons are loosely bound to the nucleus and can be treated as free, the scattered rays are found to have a smaller frequency in addition to the radiation of original frequency. This is known as the Compton Effect. The change in frequency of the incident radiation is independent of its initial frequency and depends only upon the angle of scattering. This can be satisfactorily explained on the quantum theory of light by making use of relativistic expressions for mass, energy, etc.

Consider the scattering of a photon of frequency ν_0 falling on an electron of mass m_0, in a frame of reference in which the electron is at rest. Let ν be the frequency of the scattered photon. Let **p** be the momentum of the electron after the collision and α the angle between the final and initial directions of the photon. This is shown in Fig. 3.4. Since the incident radiation

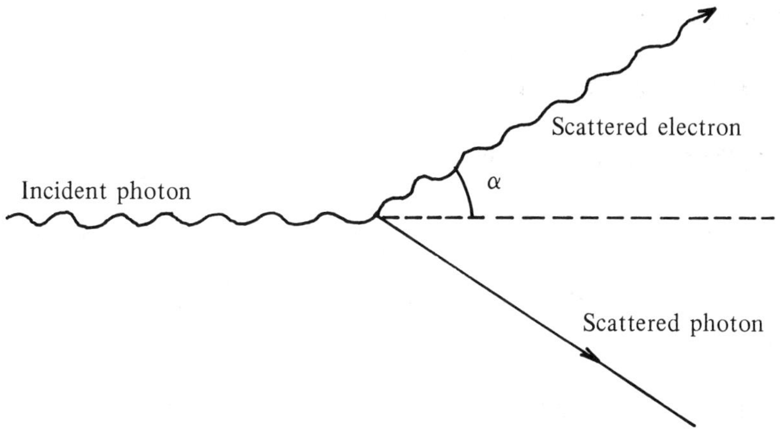

Fig. 3.4. Compton effect

is of high frequency, the incoming photon, by virtue of the relation $E = h\nu$, is very energetic. Therefore the electron which is lightly bound to its atom may be considered, to a good approximation, as free. Thus the Compton effect may be treated as a problem of collision between a photon and a stationary free electron. The initial energy of the photon is $h\nu_0$ and its initial momentum is $h\nu_0/c$, while its final energy and momentum are $h\nu$ and $h\nu/c$ respectively. According to the laws of conservation of energy and momentum, we have

$$h\,\nu_0 + m_0\,c^2 = h\,\nu + E \qquad \text{(conservation of energy)} \qquad (3.139)$$

$$\frac{h\nu_0}{c}\,\mathbf{a} = \frac{h\nu}{c}\,\mathbf{b} + \mathbf{p} \qquad \text{(conservation of momentum)} \qquad (3.140)$$

where **a** and **b** are unit vectors in the directions of the incident and the scattered photon and E is the energy of the electron after the collision. Rearranging the terms in equation (3.139), we have

$$h\nu_0 - h\nu = E - m_0 c^2.$$

Squaring the two sides, we get

$$h^2 \nu_0^2 + h^2 \nu^2 - 2 h^2 \nu_0 \nu = E^2 + m_0^2 c^4 - 2 E m_0 c^2. \quad (3.141)$$

Rearranging the terms in equation (3.140), we have

$$h\nu_0 \mathbf{a} - h\nu \mathbf{b} = c\,\mathbf{p}.$$

Squaring the two sides, we obtain

$$h^2 \nu_0^2 + h^2 \nu^2 - 2 h^2 \nu_0 \nu \cos\alpha = c^2 p^2. \quad (3.142)$$

Substituting the expression for $c^2 p^2$ from equation $E^2 = m_0^2 c^4 + p^2 c^2$ in equation (3.142), we get

$$h^2 \nu_0^2 + h^2 \nu^2 - 2 h^2 \nu_0 \nu \cos\alpha = E^2 - m_0^2 c^4. \quad (3.143)$$

Subtracting equation (3.141) from equation (3.143):

$$2 h^2 \nu_0 \nu (1 - \cos\alpha) = 2 m_0 c^2 (E - m_0 c^2) = 2 m_0 c^2 (h\nu_0 - h\nu),$$

where in obtaining the expression on the extreme right, we have made use of equation (3.139)

or $$\frac{h}{m_0 c^2}(1 - \cos\alpha) = \frac{\nu_0 - \nu}{\nu_0 \nu} = \frac{1}{\nu} - \frac{1}{\nu_0} = \frac{\lambda}{c} - \frac{\lambda_0}{c}.$$

or $$\lambda - \lambda_0 = \frac{h}{m_0 c}(1 - \cos\alpha). \quad (3.144)$$

The quantity h/m_0c is called the Compton wavelength and has the value 0.0024 nm.

Equation (3.144) has been confirmed experimentally.

Problem

Neutral pions π^0 are produced in the laboratory by bombarding stationary protons p with high energy photons of known momentum according to the reaction

$$\gamma + p \rightarrow \pi^0 + p.$$

Calculate the momentum of a pion emitted in a direction making an angle θ with the direction of the incident photon.

Solution

Fig. 3.5 illustrates the reaction $\gamma\ p \rightarrow \pi^0\ p$. According to the laws of conservation of energy and momentum, we must have

$$E_\gamma + m_p c^2 = E_\pi + E_p \tag{3.145}$$

$$\mathbf{p}_\gamma = \mathbf{p}_\pi + \mathbf{p}_p. \tag{3.146}$$

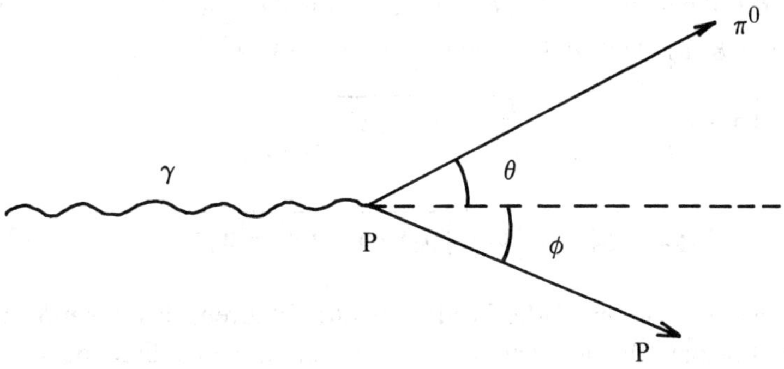

Fig. 3.5. The reaction $\gamma + p \rightarrow \pi^0 + p$

Moreover, using the formula (3.45), viz.,

$$E^2 = m_0^2 c^4 + p^2 c^2,$$

for the particles involved, we can write

$$E_\gamma^2 = p_\gamma^2 c^2 \tag{3.147a}$$

$$E_\pi^2 = p_\pi^2 c^2 + m_\pi^2 c^4 \tag{3.147b}$$

$$E_p^2 = p_p^2 c^2 + m_p^2 c^4 \tag{3.147c}$$

Substituting the expressions for E_γ, E_π and E_p from equations (3.147) in equation (3.145), we get

$$|p_\gamma| c + m_p c^2 = \sqrt{p_\pi^2 c^2 + m_\pi^2 c^4} + \sqrt{p_p^2 c^2 + m_p^2 c^4} \tag{3.148}$$

Now from equation (3.146), we obtain

$$\mathbf{p}_p = \mathbf{p}_\gamma - \mathbf{p}_\pi.$$

Squaring, we obtain

$$p_p^2 = p_\gamma^2 + p_\pi^2 - 2 |p_\gamma| |p_\pi| \cos\theta, \tag{3.149}$$

where θ is the angle which the emitted pion makes with the incident photon. Eliminating p_p^2 between equations (3.148) and (3.149), we get

$$|p_\gamma| c + m_p c^2 = \sqrt{p_\pi^2 c^2 + m_\pi^2 c^4}$$

$$+ \sqrt{(p_\gamma^2 + p_\pi^2 - 2 |p_\gamma| |p_\pi| \cos\theta) c^2 + m_p^2 c^4}\,. \tag{3.150}$$

Since the momentum of the incident photon is given, this equation can be used to calculate the momentum of the pion emitted in a direction making an angle θ with the direction of the incident photon.

Emission of a Photon from an Excited Nucleus

We shall now show that if an excited nucleus of rest mass M_0 emits a photon in a transition to the ground state and thereby loses internal energy ΔE, then the frequency ν of the photon is given by the relation

$$\nu = \frac{\Delta E}{h}\left(1 - \frac{\Delta E}{2 M_0 c^2}\right),$$

where h is Planck's constant.

Consider a frame of reference in which the excited nucleus of mass

M_0 is at rest. Let m be the mass and **V** the velocity of recoil of the nucleus when it has emitted the photon of frequency ν, and has fallen into a lower energy state. Then according to the laws of conservation of energy and momentum, we should have

$$M_0 c^2 = m c^2 + h \nu \tag{3.151}$$

$$0 = m \mathbf{V} + \frac{h\nu}{c} \mathbf{a}, \tag{3.152}$$

where **a** is a unit vector in the direction of emission of the photon. From equation (3.152), we have

$$m \mathbf{V} = -\frac{h\nu}{c} \mathbf{a}.$$

Squaring both sides of this equation, we get

$$m^2 V^2 = \frac{h^2\nu^2}{c^2}. \tag{3.153}$$

If m_0 is the rest mass of the nucleus after the emission of the photon, then

$$m = \frac{m_0}{\sqrt{1 - V^2/c^2}}$$

or $$m^2 (1 - V^2/c^2) = m_0^2 . \tag{3.154}$$

Substituting the value of v^2 from equation (3.153) in equation (3.154), we obtain

$$m^2 - \frac{h^2\nu^2}{c^4} = m_0^2 . \tag{3.155}$$

Now rearranging the terms in equation (3.151) and squaring, we get

$$(M_0 c^2 - h \nu)^2 = m^2 c^4.$$

or $$M_0^2 c^4 + h^2 \nu^2 - 2 M_0 c^2 h \nu = m^2 c^4. \tag{3.156}$$

Substituting the expression for m^2 from equation (3.155) in equation (3.152), we get

$$M_0^2 c^4 + h^2 \nu^2 - 2 M_0 c^2 h \nu = h^2 \nu^2 + m_0^2 c^4$$

or $$M_0^2 - 2 M_0 \frac{h\nu}{c^2} = m_0^2. \qquad (3.157)$$

But the change ΔE in the internal energy of the nucleus is equal to the change in the rest mass energies in its two states so that

$$\Delta E = (M_0 - m_0)\, c^2 \qquad (3.158)$$

Eliminating m_0 between equations (3.157) and (3.158), we obtain

$$M_0^2 - 2 M_0 \frac{h\nu}{c^2} = (M_0 - \frac{\Delta E}{c^2})^2 = M_0^2 + \frac{(\Delta E)^2}{c^4} - \frac{2M_0}{c^2}\Delta E$$

or $$-\frac{2M_0}{c^2} h \nu = \frac{(\Delta E)^2}{c^4} - \frac{2M_0}{c^2}\Delta E$$

or $$h \nu = \Delta E \left(1 - \frac{\Delta E}{2 M_0 c^2} \right). \qquad (3.159)$$

This is the desired result.

If ΔE is small, the second term on the right hand side of equation (3.159) may be neglected and we may write $\Delta E = h\nu$. This indicates that the term

$$\frac{(\Delta E)^2}{2 M_0 c^2}$$

represents the kinetic energy of the recoiling nucleus.

It was noted in 1957 by Mössbauer that if the emitter nucleus is bound in a crystal, e.g., ^{57}Fe, the crystal recoils as a whole. Since the mass of a crystal is very large, the effective recoil energy is negligible. Consequently, the energy spectrum of the emitted gamma rays is of extremely small width. This is known as Mössbauer effect and has been used, *inter alia*, to verify the prediction of general relativity that the frequency of electromagnetic radiation is dependent on the gravitational field.

PROBLEMS

1. The rest mass of a proton is 938.259 MeV. Calculate its momentum and kinetic energy in the laboratory when it is moving in it with a speed of 0.9 c.
2. Calculate the energy available in the laboratory frame for the production of new particles when a proton of energy 20 MeV collides with a stationary proton of rest mass m_p.
3. Find the threshold energy for the reaction $\pi^- + p \rightarrow K^+ + \Sigma^-$ in the laboratory frame of reference, the proton being stationary in the laboratory frame. What happens to the unconverted energy in the laboratory frame?
4. Calculate the threshold kinetic energy for the production of pion-antipion pair in the laboratory for the reaction

$$\pi^- + p \rightarrow n + (\pi^- + \pi^+).$$

5. Find the minimum kinetic energy required for the reaction

$$\pi^- + p \rightarrow K^+ + K^- + n$$

to be energetically possible.

6. Calculate the amount of energy released in the decay of positive kaon according to the reaction

$$K^+ \rightarrow \pi^+ + \pi^+ + \pi^-.$$

7. Show that if the process $p + \overline{p} \rightarrow p + \overline{p} + \pi^+ + \pi^-$ occurs, the threshold energy would be 0.6 GeV.
8. Show that the threshold energy required in the photon-production of neutral pi–mesons from the stationary protons in the reaction

$$\gamma + p \rightarrow p + \pi^0$$

is

$$T = M_\pi c^2 \left(1 + \frac{M_\pi}{2m_p}\right).$$

9. A neutron of rest mass m_1 moving with a uniform velocity v collides with a stationary nucleus of rest mass m_2 and is absorbed by it. Show that the rest mass M and the velocity V of the compound nucleus are given by

$$M = m_1^2 + m_1^2 + \frac{2\,m_1\,m_2}{\sqrt{1 - v^2/c^2}}$$

$$V = \frac{m_1\,v}{m_1 + m_2\sqrt{1 - v^2/c^2}}\,.$$

10. In a cloud chamber photograph, a Σ^+-hyperon was observed to decay into a proton and a neutral pion, $\Sigma^+ \rightarrow p + \pi^0$, which moved with momentum p_1 and p_2. If ϕ is the angle between the directions of motion of p and π^0, determine the rest mass of the Σ^+-hyperon.

11. A particle of non-zero rest mass moving with velocity V decays into two photons. Determine the minimum angle between the directions of motion of the photons.

12. A particle of rest mass m_0, moving with speed V in an inertial frame, collides and coalesces with a particle of rest mass m_0^*. What is the rest mass of the resulting particle?.

13. A photographic plate is exposed to the monochromatic light of wavelength 535×10^{-10} m. Calculate the number of photons falling per second on it so as to produce a force of 10^{-7} newton.

14. In the laboratory frame S, a particle of rest mass m_1, momentum $\mathbf{p}(1)$ and energy E_1 is incident on a stationary particle of rest mass m_2 and energy E_2. If starred letters denote the corresponding quantities in their CM frame S*, show that

$$E_1^* \,(E_1^* + E_2^*) = m_1^2 + E_1\,m_2$$

$$\gamma = \frac{E_1 + E_2}{E_1^* + E_2^*}$$

$$\mathbf{p}^*(1) = \frac{m_2\,\mathbf{p}(1)}{E_1^* + E_2^*}\,.$$

4

RELATIVISTIC OPTICS

In this chapter we will use the relativistic transformation laws for energy and 3-momentum to explain two optical phenomena, viz., the Doppler effect and the aberration of light.

Doppler Effect

The change in the frequency of light radiation emitted by a source due to the relative motion of the source and the observer is called the **Doppler effect** for light radiation. We will use the transformation laws for energy and momentum to derive the relativistic formula for the Doppler effect.

Let S and S$'$ be two frames of reference such that S$'$ is moving

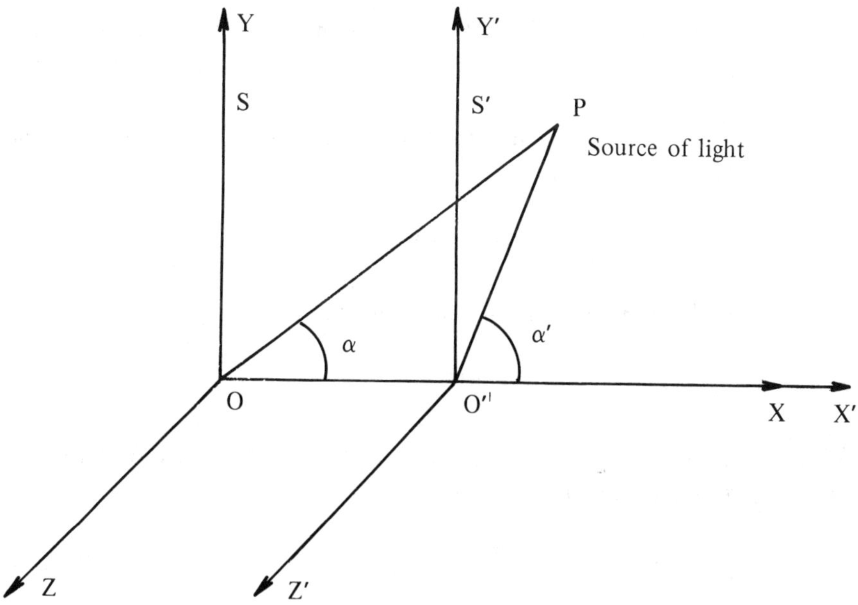

Fig. 4.1. Doppler effect

relative to S with a velocity v along the positive direction of the x-axis, with x$'$-axis coincident with it as shown in Fig. 4.1. Suppose that a source P of monochromatic radiation is at rest in S' and is emitting radiation of frequency ν' as measured by an observer at the origin O' in S'. Then, according to this observer, the energy and magnitude of the momentum of the photon would be given by $E' = h\nu'$ and $p' = h\nu'/c$, respectively. If ν is the frequency of the same radiation as measured by an observer at the origin O of S, then according to him the energy and the magnitude of the momentum of the photon would be $E = h\nu$ and $p = h\nu/c$. The relation between ν and ν' is obtained in the following way.

The transformation law for energy is given by

$$E' = \gamma\,(E - v\,p_x). \tag{3.44'}$$

The inverse transformation law is obtained by changing the sign of v and by interchanging the primed and unprimed quantities. Thus, we have

$$E = \gamma\,(E' + v\,p_x'). \tag{4.1}$$

If α' is the angle which the line $\overrightarrow{OP'}$ makes with the positive direction of the x'-axis, then

$$\begin{aligned} p_x &= \text{the x-component of the momentum of the photon in the } S' \text{ frame} \\ &= \frac{h\nu'}{c}\cos(\pi - \alpha'), \end{aligned}$$

because light is propagating in the direction $\overrightarrow{PO'}$.

or $$p_x = -\frac{h\nu'}{c}\cos\alpha'.$$

Writing $h\nu$, $h\nu'$ for E, E', and substituting the above expression for p_x' in equation (4.1) and simplifying, we get

$$\nu = \gamma\nu'\left(1 - \frac{v}{c}\cos\alpha'\right), \tag{4.2}$$

where ν' is called the **proper frequency** of light. This formula gives the frequency ν' of a light source which is moving with a velocity v relative to the observer in S. The change in frequency due to the relative motion between the source and the observer is called the relativistic Doppler shift. Note that in classical physics there are two formulae for the Doppler effect: one for the moving source and the other for the moving observer. In special relativity, we have only one formula for the Doppler effect because only the relative motion between the source and the observer is significant. It may be noticed that we need not assume that the source is moving with uniform velocity; v may be considered as the instantaneous velocity of the source.

The inverse transformation for the Doppler effect is obtained by changing the sign of v and interchanging the primed and unprimed variables:

$$\nu' = \gamma \nu \left(1 + \frac{v}{c} \cos \alpha\right), \tag{4.3}$$

where α is the angle which the line $\overrightarrow{OP}$ makes with the x-axis.

We shall now consider two special cases:

1. The Longitudinal Doppler Effect

When the source P is receding directly away from the observer O, $\alpha' = 0$ and equation (4.2) gives

$$\begin{aligned} \nu &= \gamma \nu' \left(1 - \frac{v}{c}\right) \\ &= \nu' \frac{1 - v/c}{\sqrt{1 - v^2/c^2}} \\ &= \nu' \sqrt{(1 - v/c)/(1 + v/c)}. \end{aligned} \tag{4.4}$$

Since $\sqrt{(1 - v/c)/(1 + v/c)} < 1$, the frequency ν is less than the proper frequency ν'. That is, when the source moves away from the observer, the frequency ν of light radiation is smaller than its proper frequency; frequency of light is shifted towards the red end of the spectrum. When the source is moving directly towards the observer, $\alpha' = \pi$ and equation (4.2) reduces to

$$\nu = \nu' \sqrt{(1 + v/c)/(1 - v/c)}. \tag{4.5}$$

This shows that $\nu > \nu'$, i.e., the frequency of light is shifted towards the violet end of the spectrum.

In terms of wavelengths, $\lambda = c/\nu$ and $\lambda' = c/\nu'$ so that, for $\alpha' = 0$ and π, equations (4.4) and (4.5) respectively reduce to

$$\lambda = \lambda' \sqrt{(1 + v/c)/(1 - v/c)}$$

and $$\lambda = \lambda' \sqrt{(1 - v/c)/(1 + v/c)}\,.$$

Problem

Show that to a first order in v/c, equations (4.4) and (4.5) reduce to the corresponding classical equations.

The relativistic formula for the longitudinal Doppler effect has since been confirmed experimentally.

2. *The Transverse Doppler Effect*

If $\alpha' = \pi/2$, then equation (4.2) yields $\nu = \gamma\,\nu'$. This equation shows that even if the observation is made at right angles to the direction of motion of the light source, a change in the frequency of the light radiation occurs. No such effect is predicted by the classical theory for which $\nu' = \nu$ when $\alpha' = \pi/2$. This transverse Doppler effect has been confirmed experimentally by utilizing the shift of the Mössbauer absorption line of ^{57}F when the absorber is set rotating about a fixed ^{57}Co source.

Aberration of Light

Consider two observers at the origins O and O′ of two standard frames of reference S and S′ so that they are in uniform collinear relative motion, and suppose that they are observing a monochromatic radiation emitted by a source P fixed in S′. We will show that at the time of their coincidence, they will assign different values α and α' to the angle which the light radiation reaching them makes with the line of their relative motion. This is called the **aberration of light**. The relation between the angles α and α' as measured by the two observers can be calculated as follows:

Let $p_\mu = (p_1, p_2, p_3, p_4)$ and $p_\mu' = (p_1', p_2', p_3', p_4')$ be the 4-momenta of a light quantum as measured in S and S′ respectively. Then the transformation law for 4-momentum gives (cf. equation (3.14a))

$$p_1' = \gamma\,(p_1 + i\,\frac{v}{c}\,p_4)$$

or $$-\frac{h\nu'}{c}\cos\alpha' = \gamma\left(-\frac{h\nu}{c}\cos\alpha + i\frac{v}{c}i\frac{h\nu}{c}\right)$$

or $$\nu'\cos\alpha' = \gamma\,\nu\left(\cos\alpha + \frac{v}{c}\right). \tag{4.6}$$

But we have shown in the preceding article that

$$\nu' = \gamma\,\nu\left(1 + \frac{v}{c}\cos\alpha\right). \tag{4.3'}$$

Dividing equation (4.6) by equation (4.3), we get

$$\cos\alpha' = \left(\cos\alpha + \frac{v}{c}\right)/\left(1 + \frac{v}{c}\cos\alpha\right).$$

This equation gives the relation between the angles α and α'.

The Doppler effect and the phenomenon of aberration of light are of great significance in astronomy. Observations of stellar spectra determine the rates at which stars are moving towards or away from us, while observations of red shifts in the spectra of distant galaxies indicate that the universe as a whole is expanding. The formula for the aberration of light relates the true position of a star to its observed position, the displacement being caused by the motion of the earth relative to the sun and the finiteness of the velocity of light.

PROBLEMS

1. A space probe travelling directly away from the earth contains a transmitter radiating at a constant frequency of 10^9 hz. The frequency of the signals reaching the earth can be measured with an accuracy of ± 1 hz. At what speed relative to the earth will the relativistic Doppler shift differ measurably from that expected classically?

   (*Manchester University*)

2. If the source of light and one of the observers are seen by the other observer to be moving in exactly opposite directions with equal velocities 0.6 c, and this observer also measures the wavelength of the light as 400 nm, calculate the frequency of the source and the wavelength observed by the other observer.

   (*Liverpool University*)

5

RELATIVITY AND ELECTROMAGNETISM IN VACUUM

According to the special theory of relativity, laws of physics must have the same form in all inertial frames of reference. Since Newton's laws of motion do not possess this property, mechanics had to be reformulated such that the new equations of motion are consistent with the special theory of relativity. This reformulation yields new concepts such as the variation of mass with speed and the equivalence of mass and energy. The effect of relativity on mechanics may therefore be considered revolutionary. On the other hand, Maxwell's equations, which govern electromagnetic phenomena, are already covariant under the Lorentz transformation. Special relativity, therefore, does not require any modification in these equations and consequently no new concepts are introduced. However, special relativity synthesizes the notions of electric and magnetic fields, exhibiting the physical reality as electromagnetic field, so that the influence of relativity on electromagnetism is merely to broaden our viewpoint and thus enhance our understanding of electromagnetism. In this chapter, we will study the electromagnetic phenomena in free space from this point of view.

Covariance of Maxwell's Equations

We will first prove the covariance of Maxwell's equations in vacuum by writing these equations in 4-vector form and will also take this opportunity to introduce certain 4-vectors.

Maxwell's equations in vacuum, save for the presence of electric charge, are

$$\text{curl } \mathbf{H} = \frac{1}{c}\frac{\partial \mathbf{E}}{\partial t} + \frac{4\pi}{c}\mathbf{J} \tag{5.1a}$$

$$\text{curl } \mathbf{E} = -\frac{1}{c}\frac{\partial \mathbf{H}}{\partial t} \tag{5.1b}$$

$$\text{div } \mathbf{E} = 4\pi\rho \tag{5.1c}$$

$$\text{div } \mathbf{H} = 0, \tag{5.1d}$$

where **E** and **H** are the electric and magnetic field strengths, ρ is the charge density and $\mathbf{J} = \rho\mathbf{V}$ is the current density, **V** being the velocity of charge. The system of units is Gaussian. In order to express these equations in 4-vector form, we proceed as follows. We know that equations (5.1b,d) imply the existence of vector and scalar potentials **A** and ϕ such that

$$\mathbf{H} = \text{curl } \mathbf{A} \tag{5.2a}$$

$$\mathbf{E} = -\text{ grad } \phi - \frac{1}{c}\frac{\partial \mathbf{A}}{\partial t}\ . \tag{5.2b}$$

The substitution of these expressions for **E** and **H** in the remaining two of Maxwell's four equations shows that in free space the potentials **A** and ϕ satisfy the inhomogeneous wave equations

$$\left(\nabla^2 - \frac{1}{c^2}\frac{\partial^2}{\partial t^2} \right) \mathbf{A} = -\frac{4\pi}{c}\mathbf{J} \tag{5.3a}$$

$$\left(\nabla^2 - \frac{1}{c^2}\frac{\partial^2}{\partial t^2} \right) \phi = -4\pi\rho, \tag{5.3b}$$

where **A** and ϕ are related by the **Lorentz condition**

$$\text{div } \mathbf{A} + \frac{1}{c}\frac{\partial \phi}{\partial t} = 0. \tag{5.3c}$$

Equations (5.3) are completely equivalent to Maxwell's equations (5.1).

Problem

Derive equations (5.2) and (5.3).

For a given **H**, equation (5.2a) does not determine **A** uniquely. Similarly, for given **E** and **A**, equation (5.2b) does not determine ϕ uniquely. In fact, the physically measurable fields **E** and **H** remain unaffected if the potentials **A** and ϕ are changed to ($\mathbf{A}$ + grad f) and $(\phi - \frac{1}{c}\,\partial f/\partial t)$ at the same time, f being an arbitrary scalar point function. Writing the new potentials as $\mathbf{A}^*$ and ϕ^*, we have

$$\mathbf{A}^* = \mathbf{A} + \text{grad f} \tag{5.4a}$$

and $$\phi^* = \phi - \frac{1}{c}\frac{\partial f}{\partial t} \ . \tag{5.4b}$$

Such a transformation of potentials is called a **gauge transformation.** The foregoing analysis shows that we are free to impose one scalar condition on the potentials. Different choices of the scalar function f lead to different conditions or, as it is usually called, different gauges. The condition usually imposed on $\mathbf{A}^*$ and ϕ^* is the Lorentz gauge given by equation (5.3c), in which the asterisks, not being required, have been omitted.

To express equations (5.3) in 4-vector form, we proceed step by step as follows:

First of all, we show that J_x, J_y, J_z, $ic\rho$ form a 4-vector.

Consider two standard frames of reference S and S′. Suppose that in the frame S′ a charge Q′ is at rest in a small volume $\Delta\tau'$, which, for convenience, we take as a rectangular box. Let ρ' be the charge density as measured by an observer in S′. Then

$$Q' = \rho' \, \Delta\tau'. \tag{5.5}$$

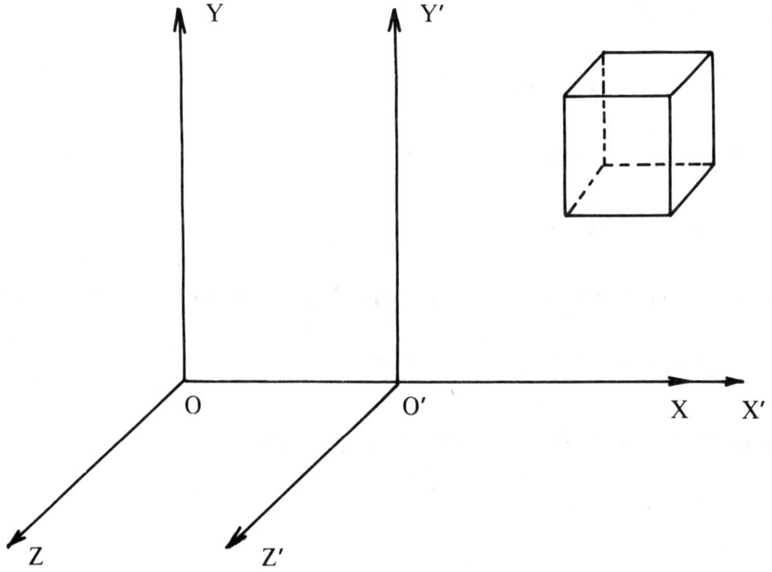

Fig.5.1. Volume contraction

Let $\Delta\tau$ be the volume, Q the charge and ρ the charge density as measured by an observer in S so that

$$Q = \rho\,\Delta\tau. \tag{5.6}$$

If we *assume* that the charge is a 4-dimensional scalar, i.e., a given charge has the same value in all inertial frames (and this is substantiated experimentally), then $Q' = Q$ and equations (5.5) and (5.6) give

$$\rho\,\Delta\tau = \rho'\,\Delta\tau'.$$

However, as $\Delta\tau = \Delta x\,\Delta y\,\Delta z$ and $\Delta\tau' = \Delta x'\,\Delta y'\,\Delta z'$, the above equation may be written as

$$\rho\,\Delta x\,\Delta y\,\Delta z = \rho'\,\Delta x'\,\Delta y'\,\Delta z'. \tag{5.7}$$

Here Δx, Δy, Δz are the lengths of the box parallel to x-, y-, z-axes as measured in S. The primed lengths correspond to the measurements in S$'$. Now as the lengths perpendicular to the direction of motion remain unchanged, we have

$$\Delta y' = \Delta y \quad \text{and} \quad \Delta z' = \Delta z,$$

and relation (5.7) reduces to

$$\rho\,\Delta x = \rho'\Delta x' \tag{5.8}$$

Moreover, as the length contracts along the direction of motion, we can write

$$\Delta x' = \gamma\,\Delta x.$$

Substituting this expression in equation (5.8), we get

$$\rho = \gamma\,\rho'. \tag{5.9}$$

Since $\gamma > 1$, the charge density ρ should be greater than ρ'; this shows that to an observer in motion relative to the charge, the charge density appears greater. The charge density ρ', measured by an observer who is at rest relative to the charge, is called the **proper charge density**.

Now since ρ' is a constant and the velocity V_μ is a 4-vector, the quantity $\rho' V_\mu$ must also be a 4-vector. We denote this 4-vector by J_μ:

$$J_\mu = \rho' V_\mu. \tag{5.10}$$

The first component of J_μ is

$$J_1 = \rho' V_1 = \rho' \frac{V_x}{\sqrt{1 - V^2/c^2}} = \gamma \rho' V_x = \rho V_x = J_x,$$

where we have used equations (3.2a, b, c) and (5.2).

Similarly, we can show that

$$J_2 = J_y,\ J_3 = J_z,$$

and $$J_4 = \rho' V_4 = \rho' \frac{1}{\sqrt{1 - V^2/c^2}}\, i\, c = \rho\, i\, c,$$

where the expression for V_4 has been substituted from equation (3.2d). Thus $(J_x, J_y, J_z, ic\rho) \equiv (J_1, J_2, J_3, J_4)$ are the components of the 4-vector J_μ:

$$J_\mu = (J_x, J_y, J_z, ic\rho) = (\mathbf{J}, ic\rho). \tag{5.11}$$

This 4-vector is called the **current density 4-vector**. Equation (5.11) shows that $\mathbf{J}$ and ρ are not independent quantities but are different components of the same entity.

Problem

Derive the equation of continuity from Maxwell's equations and write it in tensor form. What is the physical significance of this equation?

Let us next express the d'Alembertian Operator $\nabla^2 - \frac{1}{c^2}\frac{\partial^2}{\partial t^2}$ $(\equiv \Box^2)$ in a symmetrical form. We have

$$\frac{\partial}{\partial t} = \frac{\partial x_4}{\partial t}\frac{\partial}{\partial x_4} = ic\frac{\partial}{\partial x_4}.$$

Differentiating with respect to t, we get

$$\frac{\partial^2}{\partial t^2} = (ic)^2 \frac{\partial^2}{\partial x_4^2} = -c^2 \frac{\partial^2}{\partial x_4^2}.$$

Then

$$\nabla^2 - \frac{1}{c^2}\frac{\partial^2}{\partial t^2} = \frac{\partial^2}{\partial x^2} + \frac{\partial^2}{\partial y^2} + \frac{\partial^2}{\partial z^2} - \frac{1}{c^2}\frac{\partial^2}{\partial t^2}$$

$$= \frac{\partial^2}{\partial x_1^2} + \frac{\partial^2}{\partial x_2^2} + \frac{\partial^2}{\partial x_3^2} + \frac{\partial^2}{\partial x_4^2}$$

$$= \frac{\partial^2}{\partial x_\nu \partial x_\nu}.$$

Substituting this expression in equations (5.3a) and (5.3b), we get

$$\frac{\partial^2 \mathbf{A}}{\partial x_\nu \partial x_\nu} = -\frac{4\pi}{c}\mathbf{J}. \tag{5.12}$$

$$\frac{\partial^2 \phi}{\partial x_\nu \partial x_\nu} = -4\pi\rho = -\frac{4\pi}{c} c\,\rho. \tag{5.13}$$

Multiplying the last equation by $i = \sqrt{-1}$, we obtain

$$\frac{\partial^2 (i\phi)}{\partial x_\nu \partial x_\nu} = -\frac{4\pi}{c}(ic\,\rho). \tag{5.14}$$

If we define a 4-component quantity A_μ by the equation

$$A_\mu = (A_x, A_y, A_z, i\phi) \equiv (\mathbf{A}, i\phi), \tag{5.15}$$

then equations (5.12) and (5.14) may be combined as

$$\frac{\partial^2 A_\mu}{\partial x_\nu \partial x_\nu} = -\frac{4\pi}{c} J_\mu. \tag{5.16}$$

The right hand side of this equation, which is the product of a universal constant $-4\pi/c$ and the 4-vector J_μ, must be itself a 4-vector. Since the operator $\partial^2/\partial x_\nu \partial x_\nu$ is Lorentz invariant, i.e., it does not change when the frame of reference is changed, in order that equation (5.16) may be valid, the quantity A_μ must be a 4-vector. This 4-vector is called the **4-vector potential.** Subject to the Lorentz condition, equation (5.16) expresses Maxwell's equations in 4-vector form. Maxwell's equations are, therefore, Lorentz covariant. Notice that the vector and scalar potentials **A** and ϕ are really different components of the same 4-vector A_μ.

Lorentz Condition in 4-Vector Form

The Lorentz or the gauge condition is expressed by the relation

$$\text{div}\,\mathbf{A} + \frac{1}{c}\frac{\partial \phi}{\partial t} = 0. \tag{5.3c$'$}$$

Let us attempt to put it in 4-vector form. Now

$$\text{div}\,\mathbf{A} = \frac{\partial A_1}{\partial x_1} + \frac{\partial A_2}{\partial x_2} + \frac{\partial A_3}{\partial x_3}$$

and $$\frac{1}{c}\frac{\partial \phi}{\partial t} = \frac{1}{c}\frac{\partial \phi}{\partial x_4}\frac{\partial x_4}{\partial t} = i\frac{\partial \phi}{\partial x_4} = \frac{\partial}{\partial x_4}(i\phi) = \frac{\partial A_4}{\partial x_4}.$$

Substituting these expressions in equation (5.3c$'$), we get

$$\frac{\partial A_1}{\partial x_1} + \frac{\partial A_2}{\partial x_2} + \frac{\partial A_3}{\partial x_3} + \frac{\partial A_4}{\partial x_4} = 0.$$

Using the summation convention, we can write it as

$$\frac{\partial A_\mu}{\partial x_\mu} = 0. \tag{5.17}$$

This is the Lorentz condition in 4-vector form; it says that the divergence of the 4-vector A_μ is zero.

Problem

Write down the Lorentz transformation equations for A_μ and J_μ.

Field Tensor and Covariance of Maxwell's Equations

We have expressed Maxwell's equations in 4-vector form in terms of 4-potential and 4-current density. However, sometimes it is required to write Maxwell's equations in 4-tensor form in terms of **E** and **H**. Let us see how this can be accomplished. We have shown that in terms of scalar and vector potentials, the electric and magnetic fields **E** and **H** may be written as:

$$\mathbf{H} = \text{curl}\,\mathbf{A} \tag{5.2a$'$}$$

$$\mathbf{E} = -\operatorname{grad}\phi - \frac{1}{c}\frac{\partial \mathbf{A}}{\partial t}\,, \tag{5.2b$'$}$$

where **A** and ϕ are related by the Lorentz gauge.
Equation (5.2a$'$) can be written as

$$(H_1, H_2, H_3) = \begin{vmatrix} \mathbf{i}_1 & \mathbf{i}_2 & \mathbf{i}_3 \\ \dfrac{\partial}{\partial x_1} & \dfrac{\partial}{\partial x_2} & \dfrac{\partial}{\partial x_3} \\ A_1 & A_2 & A_3 \end{vmatrix}.$$

Equating the coefficients of $\mathbf{i}_1$, $\mathbf{i}_2$, $\mathbf{i}_3$, we get

$$\begin{aligned} H_1 &= \frac{\partial A_3}{\partial x_2} - \frac{\partial A_2}{\partial x_3} \\ H_2 &= \frac{\partial A_1}{\partial x_3} - \frac{\partial A_3}{\partial x_1} \\ H_3 &= \frac{\partial A_2}{\partial x_1} - \frac{\partial A_1}{\partial x_2}\,. \end{aligned} \tag{5.18}$$

Moreover, equation (5.2b$'$) gives

$$\begin{aligned} (E_1, E_2, E_3) &= -\left(\frac{\partial \phi}{\partial x_1}, \frac{\partial \phi}{\partial x_2}, \frac{\partial \phi}{\partial x_3}\right) - \frac{1}{c}\left(\frac{\partial A_x}{\partial t}, \frac{\partial A_y}{\partial t}, \frac{\partial A_z}{\partial t}\right) \\ &= i\left(\frac{\partial A_4}{\partial x_1}, \frac{\partial A_4}{\partial x_2}, \frac{\partial A_4}{\partial x_3}\right) - \frac{1}{c}\left(\frac{\partial A_1}{\partial x_4}, \frac{\partial A_2}{\partial x_4}, \frac{\partial A_3}{\partial x_4}\right), \end{aligned}$$

because $A_4 = i\,\phi$. Comparing the coefficients of $\mathbf{i}_1$, $\mathbf{i}_2$, $\mathbf{i}_3$ on the two sides of this equation, we get

$$\begin{aligned} -i\,E_1 &= \frac{\partial A_4}{\partial x_1} - \frac{\partial A_1}{\partial x_4} \\ -i\,E_2 &= \frac{\partial A_4}{\partial x_2} - \frac{\partial A_2}{\partial x_4} \\ -i\,E_3 &= \frac{\partial A_4}{\partial x_3} - \frac{\partial A_3}{\partial x_4}\,. \end{aligned} \tag{5.19}$$

The right hand sides of equations (5.18) and (5.19) are of the form of the components of the curl of a vector. We know that a curl operation on a vector really defines an antisymmetric tensor of rank 2. These equations therefore suggest that Maxwell's equations which involve electric and magnetic fields may be expressed in terms of an antisymmetric tensor of rank 2.

We now define a 4-dimensional antisymmetric second-rank tensor $f_{\mu\nu}$ by the equation

$$f_{\mu\nu} = \frac{\partial A_\nu}{\partial x_\mu} - \frac{\partial A_\mu}{\partial x_\nu} . \qquad (5.20)$$

For $\mu = 1,\ \nu = 2$, this gives

$$f_{12} = \frac{\partial A_2}{\partial x_1} - \frac{\partial A_1}{\partial x_2} = H_3. \qquad \text{(by equation (5.18))}$$

For $\mu = 1,\ \nu = 3$, we get

$$f_{13} = \frac{\partial A_3}{\partial x_1} - \frac{\partial A_1}{\partial x_3} = -H_2. \qquad \text{(by equation (5.18))}$$

Similary, we obtain

$$f_{23} = H_1,\ \ f_{14} = -i\,E_1,\ \ f_{24} = -i\,E_2,\ \ f_{34} = -i\,E_3.$$

Since $f_{\mu\nu}$ is antisymmetric:

$$f_{\mu\nu} = 0, \ \text{ if } \mu = \nu$$

and $$f_{\mu\nu} = -f_{\nu\mu}, \ \text{ if } \mu \neq \nu.$$

All the 16 components of the tensor $f_{\mu\nu}$ can now be arranged as a matrix:

$$f_{\mu\nu} = \begin{pmatrix} 0 & H_3 & -H_2 & -i\,E_1 \\ -H_3 & 0 & H_1 & -i\,E_2 \\ H_2 & -H_1 & 0 & -i\,E_3 \\ i\,E_1 & i\,E_2 & i\,E_3 & 0 \end{pmatrix} . \qquad (5.21)$$

The 4-tensor $f_{\mu\nu}$ is called the **electromagnetic field tensor**. This shows that the electromagnetic field is an antisymmetric tensor of rank 2 in 4-dimensional space, *uniting electric and magnetic fields*. Thus, in special relativity, the electric and magnetic fields do not exist as independent entities; instead, the physical concept emerges as the *electromagnetic field*.

We are now well equipped to write down Maxwell's equations in 4-vector form. First consider the equation

$$\text{div}\,\mathbf{E} = 4\pi\rho.$$

It can be written as

$$\frac{\partial E_1}{\partial x_1} + \frac{\partial E_2}{\partial x_2} + \frac{\partial E_3}{\partial x_3} = 4\pi\rho.$$

Substituting the expressions for E_1, E_2, E_3 in terms of the components of the field tensor $f_{\mu\nu}$ and writing J_4 for $ic\rho$, we have

$$\frac{\partial f_{41}}{\partial x_1} + \frac{\partial f_{42}}{\partial x_2} + \frac{\partial f_{43}}{\partial x_3} = \frac{4\pi}{c} J_4.$$

Since $f_{44} = 0$, we can add $\partial f_{44}/\partial x_4$ to the left hand side without affecting its value:

$$\frac{\partial f_{41}}{\partial x_1} + \frac{\partial f_{42}}{\partial x_2} + \frac{\partial f_{43}}{\partial x_3} + \frac{\partial f_{44}}{\partial x_4} = \frac{4\pi}{c} J_4.$$

Using the summation convention, we can write it as

$$\frac{\partial f_{4\nu}}{\partial x_\nu} = \frac{4\pi}{c} J_4. \qquad (5.22)$$

Let us next consider equation

$$\text{curl}\,\mathbf{H} = \frac{1}{c}\frac{\partial \mathbf{E}}{\partial t} + \frac{4\pi}{c}\mathbf{J}.$$

We can write it as

$$\begin{vmatrix} \mathbf{i}_1 & \mathbf{i}_2 & \mathbf{i}_3 \\ \dfrac{\partial}{\partial x_1} & \dfrac{\partial}{\partial x_2} & \dfrac{\partial}{\partial x_3} \\ H_1 & H_2 & H_3 \end{vmatrix} = \frac{1}{c}\frac{\partial \mathbf{E}}{\partial x_4}\frac{\partial x_4}{\partial t} + \frac{4\pi}{c}\mathbf{J}$$

$$= i \frac{\partial}{\partial x_4} (E_1, E_2, E_3) + \frac{4\pi}{c} (J_1, J_2, J_3).$$

Equating the coefficients of $\mathbf{i}_1$, we get

$$\frac{\partial H_3}{\partial x_2} - \frac{\partial H_2}{\partial x_3} = \frac{\partial}{\partial x_4} (i\, E_1) + \frac{4\pi}{c} J_1.$$

Substituting the expressions for H_3, H_2 and E_1 from equation (5.21) and noting that $f_{11} = 0$, we get

$$\frac{\partial f_{11}}{\partial x_1} + \frac{\partial f_{12}}{\partial x_2} + \frac{\partial f_{13}}{\partial x_3} + \frac{\partial f_{14}}{\partial x_4} = \frac{4\pi}{c} J_1$$

or
$$\frac{\partial f_{1\nu}}{\partial x_\nu} = \frac{4\pi}{c} J_1. \tag{5.23a}$$

Similarly, equating the coefficients of $\mathbf{i}_2$ and $\mathbf{i}_3$ and simplifying, we obtain

$$\frac{\partial f_{2\nu}}{\partial x_\nu} = \frac{4\pi}{c} J_2 \tag{5.23b}$$

$$\frac{\partial f_{3\nu}}{\partial x_\nu} = \frac{4\pi}{c} J_3. \tag{5.23c}$$

Combining equations (5.22) and (5.23), we get

$$\frac{\partial f_{\mu\nu}}{\partial x_\nu} = \frac{4\pi}{c} J_\mu. \tag{5.24}$$

This equation therefore represents two of Maxwell's four equations.

Let us next consider the remaining two equations of Maxwell. The equation

$$\text{div}\, \mathbf{H} = 0$$

gives
$$\frac{\partial H_1}{\partial x_1} + \frac{\partial H_2}{\partial x_2} + \frac{\partial H_2}{\partial x_2} = 0.$$

Substituting the expressions for H_1, H_2, H_3 in terms of the components of the electromagnetic field tensor $f_{\mu\nu}$, the above equation reduces to

$$\frac{\partial f_{23}}{\partial x_1} + \frac{\partial f_{31}}{\partial x_2} + \frac{\partial f_{12}}{\partial x_3} = 0. \tag{5.25}$$

The last equation

$$\text{curl } \mathbf{E} = -\frac{1}{c}\frac{\partial \mathbf{H}}{\partial t}$$

can be written as

$$\begin{vmatrix} \mathbf{i}_1 & \mathbf{i}_2 & \mathbf{i}_3 \\ \dfrac{\partial}{\partial x_1} & \dfrac{\partial}{\partial x_2} & \dfrac{\partial}{\partial x_3} \\ E_1 & E_2 & E_3 \end{vmatrix} = \frac{1}{c}\frac{\partial \mathbf{H}}{\partial x_4}\frac{\partial x_4}{\partial t}$$

$$= -i\frac{\partial \mathbf{H}}{\partial x_4} = -i\frac{\partial}{\partial x_4}(H_1, H_2, H_3).$$

Equating the coefficients of $\mathbf{i}_1$ on the two sides of this equation, we get

$$\frac{\partial E_3}{\partial x_2} - \frac{\partial E_2}{\partial x_3} = -i\frac{\partial H_1}{\partial x_4}.$$

With the help of equation (5.21), this can be reduced to

$$\frac{\partial f_{43}}{\partial x_2} + \frac{\partial f_{24}}{\partial x_3} + \frac{\partial f_{32}}{\partial x_4} = 0. \tag{5.26a}$$

Similarly, we can show that

$$\frac{\partial f_{14}}{\partial x_3} + \frac{\partial f_{31}}{\partial x_4} + \frac{\partial f_{42}}{\partial x_1} = 0. \tag{5.26b}$$

and

$$\frac{\partial f_{21}}{\partial x_4} + \frac{\partial f_{42}}{\partial x_1} + \frac{\partial f_{14}}{\partial x_2} = 0. \tag{5.26c}$$

Combining equations (5.25) and (5.26), we get

$$\frac{\partial f_{\mu\nu}}{\partial x_\sigma} + \frac{\partial f_{\nu\sigma}}{\partial x_\mu} + \frac{\partial f_{\sigma\mu}}{\partial x_\nu} = 0. \tag{5.27}$$

which represents the remaining two equations of Maxwell.

Hence the equations

$$\frac{\partial f_{\mu\nu}}{\partial x_\nu} = \frac{4\pi}{c} J_\mu \tag{5.28}$$

$$\frac{\partial f_{\mu\nu}}{\partial x_\sigma} + \frac{\partial f_{\nu\sigma}}{\partial x_\mu} + \frac{\partial f_{\sigma\mu}}{\partial x_\nu} = 0 \tag{5.29}$$

are Maxwell's equations in 4-tensor form, showing that Maxwell's equations are Lorentz covariant.

Equations (5.28) are equivalent to equations (5.1a,c) which when written in scalar form are four in number and therefore are said to represent the **Faraday tetrad.** Equations (5.29) are equivalent to equations (5.1b,d) and are said to represent the **Ampere-Maxwell tetrad.**

If we use the notation

$$f_{\mu\nu,\sigma} = \frac{\partial f_{\mu\nu}}{\partial x_\sigma},$$

the above equations may be written as

$$f_{\mu\nu,\nu} = \frac{4\pi}{c} J_\mu \tag{5.30}$$

$$f_{\mu\nu,\sigma} + f_{\nu\sigma,\mu} + f_{\sigma\mu,\nu} = 0. \tag{5.31}$$

Transformation Equations for E and H

Let us now find out how the components of the electric and magnetic fields **E** and **H** are transformed as we go from one standard frame of reference S to another S$'$, i.e., given **E** and **H** in one frame, how do they look in another frame moving past? To determine this we proceed as follows:

We know that the components of the electromagnetic field constitute the field tensor $f_{\mu\nu}$ as given by the equation

$$f_{\mu\nu} = \begin{pmatrix} 0 & H_3 & -H_2 & -iE_1 \\ -H_3 & 0 & H_1 & -iE_2 \\ H_2 & -H_1 & 0 & -iE_3 \\ iE_1 & iE_2 & iE_3 & 0 \end{pmatrix} \tag{5.21$'$}$$

But $f_{\mu\nu}$ is a second-rank tensor. Its components therefore must transform as

$$f'_{\mu\nu} = a_{\mu\alpha}\, a_{\nu\beta}\, f_{\alpha\beta}. \qquad (5.32)$$

The values of $f_{\alpha\beta}$ and $f'_{\mu\nu}$ in terms of the components of unprimed and primed electric and magnetic fields can be read from equation (5.21) and a similar equation for primed quantities. The values of a's can be read from equation (2.19). Hence the transformation laws for **E** and **H**.

To determine these laws, we first consider the component f'_{23}, for which $\mu = 2$, $\nu = 3$. This component transforms as

$$f'_{23} = a_{2\alpha}\, a_{3\beta}\, f_{\alpha\beta}.$$

As α and β are dummy indices, the summation must be carried over them, the range of values being from 1 to 4. However, from equation (2.19), the only non-zero a's occurring in this equation are a_{22} and a_{33}. Therefore, summing over α, we obtain

$$f'_{23} = a_{22}\, a_{3\beta}\, f_{2\beta}.$$

Next summing over β, we get

$$f'_{23} = a_{22}\, a_{33}\, f_{23}.$$

Since, by equation (2.19), $a_{22} = a_{33} = 1$, the above equation reduces to

$$f'_{23} = f_{23}. \qquad (5.33)$$

Substituting the expression for f_{23} from equation (5.21), and writing a similar expression for f'_{23}, we obtain

$$H'_1 = H_1. \qquad (5.34a)$$

This equation shows that the first component of the magnetic field **H** is an invariant quantity.

Let us next consider the case when $\mu = 3$, $\nu = 1$. Then equation (5.32) gives

$$f'_{31} = a_{3\alpha}\, a_{1\beta}\, f_{\alpha\beta}.$$

From equation (2.19), the only non-zero a's occurring in this equation are a_{33}, a_{11}, a_{14} and are equal to 1, γ, $iv\gamma/c$ respectively. Therefore, summing over α and β, we get

$$f'_{31} = a_{33}\, a_{11}\, f_{31} + a_{33}\, a_{14}\, f_{34} = \gamma\, f_{31} + \frac{i\, v\, \gamma}{c}\, f_{34}$$

$$= \gamma\, (f_{31} + \frac{i\, v}{c}\, f_{34}).$$

Substituting the expressions for f'_{31}, f_{31} and f_{34} in the above equation, by using equation (5.21) and a similar relation for the primed quantities, we get

$$H'_2 = \gamma\, [H_2 + \frac{i\, v}{c}\, (-i\, E_3)] = \gamma\, (H_2 + \frac{v}{c}\, E_3). \qquad (5.34b)$$

This equation shows how the second component of the magnetic field in the primed frame of reference is related to the components of the electric and magnetic fields in the unprimed frame.

Now for $\mu = 1$, $\nu = 2$, we have

$$f'_{12} = a_{1\alpha}\, a_{2\beta}\, f_{\alpha\beta}.$$

Summing over α and β, and noting that

$$a_{11} = \gamma, \quad a_{14} = \frac{i\, v\, \gamma}{c}, \quad a_{22} = 1$$

while the rest of the a's are zero, we obtain

$$f'_{12} = a_{11}\, a_{22}\, f_{12} + a_{14}\, a_{22}\, f_{42} = \gamma\, f_{12} + \frac{i\, v\, \gamma}{c}\, f_{42}.$$

By virtue of the field tensor, we obtain

$$H'_3 = \gamma\, (H_3 - \frac{v}{c}\, E_2). \qquad (5.34c)$$

Similarly, by giving different values to μ and ν, we can show that

$$E'_1 = E_1 \qquad (5.35a)$$

$$E_2' = \gamma \left(E_2 - \frac{v}{c} H_3\right). \tag{5.35b}$$

$$E_3' = \gamma \left(E_3 + \frac{v}{c} H_2\right). \tag{5.35c}$$

Equations (5.34) and (5.35) give the transformation laws for **E** and **H**. They reveal the close interconnection between the electric and magnetic fields. It follows from these equations that the electric *and* magnetic fields do not have independent existence. A pure electric field in the frame S (**E** $\neq 0$, **H** $= 0$) transforms into electric and magnetic fields in S′. However, it is impossible to find a Lorentz frame such that a pure electric field in S transforms into a pure magnetic field in that frame.

We summarize below the results on the relativistic transformations for the electromagnetic field:

$$E_1' = E_1$$

$$E_2' = \gamma \left(E_2 - \frac{v}{c} H_3\right).$$

$$E_3' = \gamma \left(E_3 + \frac{v}{c} H_2\right).$$

$$H_1' = H_1.$$

$$H_2' = \gamma \left(H_2 + \frac{v}{c} E_3\right).$$

$$H_3' = \gamma \left(H_3 - \frac{v}{c} E_2\right).$$

The transformations can be written more simply if we denote the field components parallel and perpendicular to the direction of the relative velocity of S and S′ by E_L, B_L and E_T, B_T, respectively. Then the transformation equations for the electromagnetic field may be written as

$$E_L' = E_L$$

$$E_T' = \gamma \left[E_T + \frac{1}{c} (\mathbf{v} \wedge \mathbf{H})_T\right]$$

$$H_L' = H_L$$

$$H'_T = \gamma\,[H_T - \frac{1}{c}(v \wedge E)_T].$$

The inverse transformations can be obtained by merely writing $-\mathbf{v}$ for $\mathbf{v}$ and interchanging the primed and unprimed variables.

Invariants of the Electromagnetic Field

An invariant quantity is that which has the same value in all frames of reference. Invariants are of special interest because of their objective nature and it is therefore always instructive to discover invariant quantities. We will now show that $H^2 - E^2$ is an invariant of the electromagnetic field.

Since $f_{\mu\nu}$ is a 4-tensor, its scalar product with itself must be an invariant quantity. Now

$$f_{\mu\nu}\,f_{\mu\nu} = f_{1\nu}\,f_{1\nu} + f_{2\nu}\,f_{2\nu} + f_{3\nu}\,f_{3\nu} + f_{4\nu}\,f_{4\nu}\,. \qquad (5.36)$$

But $$f_{1\nu}\,f_{1\nu} = f_{11}^2 + f_{12}^2 + f_{13}^2 + f_{14}^2\,.$$

Inserting the values of the components f_{11}, etc. from equation (5.21), we get

$$f_{1\nu}\,f_{1\nu} = H_3^2 + H_2^2 - E_1^2.$$

Similarly, we can prove that

$$f_{2\nu}\,f_{2\nu} = H_1^2 + H_3^2 - E_2^2$$

$$f_{3\nu}\,f_{3\nu} = H_2^2 + H_1^2 - E_3^2$$

$$f_{4\nu}\,f_{4\nu} = -\;E_1^2 - E_2^2 - E_3^2.$$

Substituting these expressions in equation (5.36), we get

$$f_{\mu\nu}\,f_{\mu\nu} = 2\,(\,H_1^2 + H_2^2 + H_3^2\,) - 2\,(\,E_1^2 + E_2^2 + E_3^2\,)$$

$$= 2\,(\mathbf{H}^2 - \mathbf{E}^2). \qquad (5.37)$$

Hence $\mathbf{H}^2 - \mathbf{E}^2$ is an invariant of the electromagnetic field.

Problem

Show that **E** • **H** is another invariant of the electromagnetic field.

It can be shown that **E** • **H** and $H^2 - E^2$ are the only independent invariants of the electromagnetic field.

For a plane electromagnetic wave, **E** • **H** = 0 and $H^2 - E^2 = 0$. Therefore, the invariance of the quantities **E** • **H** and $H^2 - E^2$ guarantees the covariance of the above equations in all inertial frames of reference.

Lorentz Force on a Charged Particle Moving in an Electromagnetic Field

We shall now calculate the mechanical force due to an electromagnetic field as experienced by a particle carrying a charge e and moving in a frame S with a velocity **V**, which may or may not be uniform. Let S′ be the instantaneous rest frame of the particle, and consider the product of the electromagnetic field tensor $f'_{\mu\nu}$ and the 4-velocity V'_ν in this frame. This product can be expanded as

$$f'_{\mu\nu} V'_\nu = f'_{\mu 1} V'_1 + f'_{\mu 2} V'_2 + f'_{\mu 3} V'_3 + f'_{\mu 4} V'_4 . \qquad (5.38)$$

Since the charged particle is at rest in the frame S′, its 4-velocity in this frame is given by

$$(V'_1, V'_2, V'_3, V'_4) = \gamma(V)\,(\mathbf{V}, i\,c) = (\mathbf{0}, i\,c), \qquad (3.2e')$$

i.e., the components of the ordinary velocity vanish while the fourth component, V'_4, of the 4-velocity is ic. Substituting these values in equation (5.38), we get

$$f'_{\mu\nu} V'_\nu = i\,c\,f'_{\mu 4}. \qquad (5.39a)$$

For $\mu = 1, 2, 3$, using the conventional notation we write j for μ:

$$f'_{j\nu} V'_\nu = i\,c\,f'_{j4} .$$

Substituting the value of f'_{j4} from an equation for the primed quantities similar to equation (5.21), we get

$$f'_{j\nu}\ V'_{\nu} = c\ E'_j . \tag{5.39b}$$

Now in the instantaneous rest frame of the particle, only the electric field exerts a force on the charged particle. This force is given by

$$\mathbf{f}' = e\ \mathbf{E}',$$

where $\mathbf{E}'$ is the strength of the electric field. In terms of the components of the vector quantities, the above equation may be written as

$$f'_j = e\ E'_j .$$

Substituting the value of E'_j from this equation in equation (5.39b), we obtain

$$f'_{j\nu}\ V'_{\nu} = \frac{c}{e}\ f'_j . \tag{5.39c}$$

But, by virtue of equation (3.23), for $\mathbf{V} = 0$, we get

$$f'_j = F'_j .$$

Therefore equation (5.39c) can be written as

$$f'_{j\nu}\ V'_{\nu} = \frac{c}{e}\ F'_j . \tag{5.39d}$$

For $\mu = 4$, equation (5.39a) gives

$$f'_{4\nu}\ V'_{\nu} = i\ c\ f'_{44} = 0,$$

because, by virtue of equation (5.21), $f'_{44} = 0$.
Moreover, as $F'_4 = 0$ for $\mathbf{V} = 0$, we may write this equation as

$$f'_{4\nu}\ V'_{\nu} = \frac{c}{e}\ F'_4 . \tag{5.39e}$$

Combining equations (5.39d) and (5.39e), we obtain

$$f'_{\mu\nu}\ V'_{\nu} = \frac{c}{e}\ F'_{\mu} .$$

Since this is a tensor equation, it must have the same form in all inertial frames of reference. Hence even in the inertial frame S where the particle is in motion, we can write

$$f_{\mu\nu} V_\nu = \frac{c}{e} F_\mu.$$

or $$F_\mu = \frac{e}{c} f_{\mu\nu} V_\nu. \tag{5.40}$$

Let us next find the explicit expressions for the components of F_μ.
For $\mu = 1$, equation (5.40) gives

$$F_1 = \frac{e}{c} f_{1\nu} V_\nu = \frac{e}{c} (f_{11} V_1 + f_{12} V_2 + f_{13} V_3 + f_{14} V_4).$$

Substituting in the above equation the expressions for F_1, $f_{\mu\nu}$ and V_μ from equations (3.17a) and (3.19a), (5.21) and (3.2), we obtain

$$\frac{f_x}{\sqrt{1 - V^2/c^2}} = \frac{e}{c}\left[H_3 \frac{V_y}{\sqrt{1 - V^2/c^2}} - H_2 \frac{V_z}{\sqrt{1 - V^2/c^2}} - i E_1 \frac{i c}{\sqrt{1 - V^2/c^2}} \right].$$

or $$f_x = e E_1 + \frac{e}{c} (V_y H_3 - V_z H_2).$$

Similarly, we can show that

$$f_y = e E_2 + \frac{e}{c} (V_z H_1 - V_x H_3).$$

$$f_z = e E_3 + \frac{e}{c} (V_x H_2 - V_y H_1).$$

These three equations can be expressed in vector form as

$$\mathbf{f} = e \mathbf{E} + \frac{e}{c} \mathbf{V} \wedge \mathbf{H}. \tag{5.41}$$

The force **f** is called the Lorentz force; it is the force experienced by a particle of charge e moving in an electromagnetic field (**E**, **H**) in a frame where the velocity of the charged particle is **V**.

Notice that the equation $\mathbf{f} = e\mathbf{E} + \frac{e}{c}\mathbf{V} \wedge \mathbf{H}$ cannot be derived in classical theory and is taken there as an additional postulate besides Maxwell's four equations. In the theory of special relativity, however, it appears quite naturally.

For $\mu = 4$, equation (5.39) gives

$$F_4 = \frac{e}{c}(f_{41} V_1 + f_{42} V_2 + f_{43} V_3 + f_{44} V_4).$$

Substituting the expressions for F_4, $f_{\mu\nu}$ and V_μ in the above equation, we get

$$\frac{1}{\sqrt{1 - V^2/c^2}} \frac{e}{c} i \mathbf{E} \cdot \mathbf{V} = \frac{e}{c}\left[i E_1 \frac{V_x}{\sqrt{1 - V^2/c^2}} + i E_2 \frac{V_y}{\sqrt{1 - V^2/c^2}} + i E_3 \frac{V_z}{\sqrt{1 - V^2/c^2}}\right]$$

$$= \frac{1}{\sqrt{1 - V^2/c^2}} \frac{e}{c} i \mathbf{E} \cdot \mathbf{V},$$

which is merely an identity and therefore does not provide any new information.

Electromagnetic Field Produced by a Uniformly Moving Point Charge

There are many cases where charged particles move with uniform and high velocity. For instance, cosmic rays coming from outer space move with a velocity which is approximately equal to that of light. It is, therefore, instructive to see what the fields produced by such particles actually look like.

To compute the field of such a rapidly moving charged particle in any frame S, we first consider the field produced by it in the inertial frame S′ in which it is at rest at the origin as shown in Fig. 5.2. There the field is merely the static electric field of a point charge. The transformation equations for the electric and magnetic fields can then be used to obtain the electromagnetic field produced by this charged particle in the frame S.

Suppose that the particle of charge e is moving with a *uniform velocity* **v** in the frame S, the x-axis of the frame being taken along the direction of motion of the charged particle as shown in Fig. 5.2. Since the orientation of any one axis of S′ is arbitrary, the x- and x′-axes may be taken as coincident. Then the electromagnetic field at the same point but in the frame S, may be obtained by using the inverse transformation equations (5.34, 5.35) for the magnetic and electric fields. It may be noted that as the charged particle moving with velocity **v** in S is at rest in S′, the frame S′ is also moving with the same velocity relative to S. It is for this reason that we have represented the velocity of the charged particle by **v**.

Now according to Coulomb's law, a charge e at rest at the origin of an inertial frame S' will produce a spherically symmetric electric field, whose strength at a point $P(x', y', z')$ is given by

$$\mathbf{E}' = \frac{e}{r'^3}\mathbf{r}', \tag{5.42a}$$

where $\mathbf{r}'$ is the position vector of P, and

$$r'^2 = x'^2 + y'^2 + z'^2. \tag{5.42b}$$

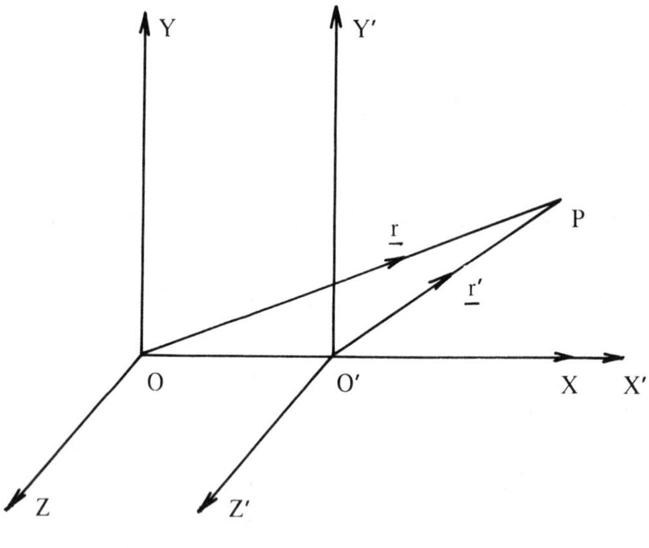

Fig. 5.2. Position of the charged particle in primed and unprimed frames

In terms of the components of $\mathbf{E}'$ and $\mathbf{r}'$, equation (5.42a) can be written as

$$(E_1', E_2', E_3') = \frac{e}{r'^3}(x', y', z'). \tag{5.43}$$

Moreover, the charge e, being at rest in S', does not produce any magnetic field and, therefore,

$$\mathbf{H}' = 0 \quad \text{or} \quad H_1' = H_2' = H_3' = 0. \tag{5.44}$$

By virtue of equations (5.43) and (5.44) and the transformation laws for the components of **E** and **H** as obtained by the inverse of transformation equations (5.34) and (5.35), the components of the electric and magnetic fields, as observed in S, can be written as

$$E_1 = E_1' = \frac{e}{r'^3}\,x' \tag{5.45a}$$

$$E_2 = \gamma\left(E_2' + \frac{v}{c}H_3'\right) = \gamma\frac{e}{r'^3}\,y' \tag{5.45b}$$

$$E_3 = \gamma\left(E_3' + \frac{v}{c}H_2'\right) = \gamma\frac{e}{r'^3}\,z' \tag{5.45c}$$

$$H_1 = H_1' = 0 \tag{5.46a}$$

$$H_2 = \gamma\left(H_2' - \frac{v}{c}E_3'\right) = -\gamma\frac{v}{c}\frac{e}{r'^3}\,z' \tag{5.46b}$$

$$H_3 = \gamma\left(H_3' + \frac{v}{c}E_2'\right) = \gamma\frac{v}{c}\frac{e}{r'^3}\,y'. \tag{5.46c}$$

This again shows that a purely electrostatic field produced by a charge in its rest frame is changed into electric and magnetic fields when viewed from a frame of reference which is in motion relative to the charge.

Equations (5.45) and (5.46) can be put in vector form as:

$$\mathbf{E} = (E_1, E_2, E_3) = \frac{e}{r'^3}(x', \gamma\, y', \gamma\, z') \tag{5.47}$$

and

$$\mathbf{H} = (H_1, H_2, H_3) = \gamma\frac{v}{c}\frac{e}{r'^3}(0, -z', y'). \tag{5.48}$$

These equations give the electromagnetic field produced by a rapidly moving charge. They show that the fields **E** and **H** are perpendicular to each other:

$$\mathbf{E} \cdot \mathbf{H} = 0.$$

In order to express **E** and **H** in terms of S-coordinates, we proceed as follows: Let (x, y, z) be the coordinates and **r** the position vector of the point P in S. Then the Lorentz transformation equations for space coordinates give

$$x' = \gamma\,(x - v\,t), \quad y' = y, \quad z' = z$$

so that

$$r'^2 = x'^2 + y'^2 + z'^2 = \gamma^2 (x - v\,t)^2 + y^2 + z^2$$

or $$r' = \sqrt{\gamma^2 (x - v\,t)^2 + y^2 + z^2}\ .$$

Substituting these expressions for x', y', z' and r' in equations (5.47) and (5.48), we get

$$\mathbf{E} = \frac{e}{[\,\gamma^2 (x - v\,t)^2 + y^2 + z^2\,]^{3/2}}[\gamma\,(x - v\,t),\ \gamma\,y,\ \gamma\,z]$$

$$= \frac{e}{[\,\gamma^2 (x - v\,t)^2 + y^2 + z^2\,]^{3/2}}\gamma\,(\,x - v\,t,\ y,\ z\,) \tag{5.49}$$

and $$\mathbf{H} = \frac{e}{c[\gamma^2 (x - v\,t)^2 + y^2 + z^2\,]^{3/2}}(0,\ -z,\ y) \tag{5.50}$$

Note that the above equations for the electromagnetic field are only valid for a uniformly moving charge and not for an accelerated charge because the acceleration of the charge does affect the field. In particular, for $t = t' = 0$, we have

$$\mathbf{E} = \frac{e\,\gamma}{[\,\gamma^2 x^2 + y^2 + z^2\,]^{3/2}}(x,\ y,\ z)$$

$$= \frac{e\,\gamma}{[\,\gamma^2 x^2 + y^2 + z^2\,]^{3/2}}\mathbf{r} \tag{5.51a}$$

$$\mathbf{H} = \frac{e\,\gamma\,v}{c\,[\,\gamma^2 x^2 + y^2 + z^2\,]^{3/2}}\ (0,\ -z,\ y)$$

$$= \frac{e\,\gamma\,v}{c\,[\,\gamma^2 x^2 + y^2 + z^2\,]^{3/2}}\mathbf{v} \wedge \mathbf{r}. \tag{5.51b}$$

These relations can be put in a more familiar form if we remember that in the frame of reference S we have

$$r^2 = x^2 + y^2 + z^2 \quad \text{and} \quad x = r\cos\theta,$$

where θ is the angle which the vector **r** makes with the x-axis. Then

$$x^2 = r^2 \cos^2\theta \quad \text{and} \quad y^2 + z^2 = r^2 - r^2\cos^2\theta = r^2\sin^2\theta \ .$$

Substituting these expressions for x^2 and $y^2 + z^2$ in equation (5.51a), we find that the expression for the electric field reduces to

$$\mathbf{E} = \frac{e\,\gamma\,\mathbf{r}}{(\gamma^2 r^2\cos^2\theta + r^2\sin^2\theta)^{3/2}}$$

$$= \frac{e\,\gamma\,\mathbf{r}}{r^3[\,\gamma^2 + (1-\gamma^2)\sin^2\theta]^{3/2}} \ . \qquad (5.52)$$

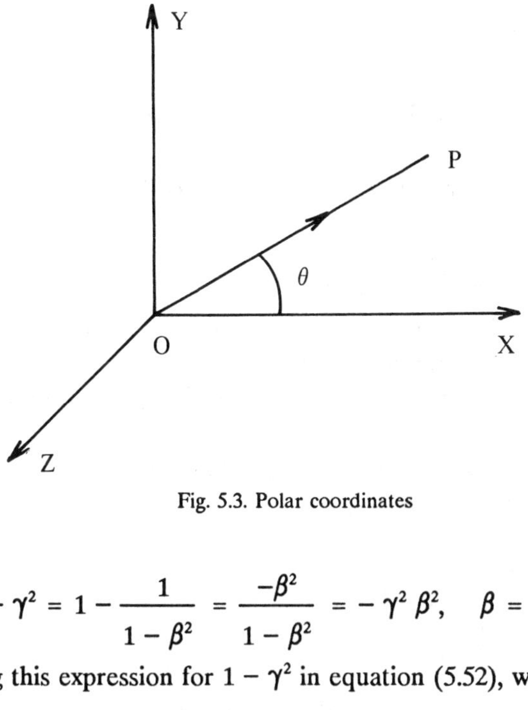

Fig. 5.3. Polar coordinates

Now $\quad 1 - \gamma^2 = 1 - \dfrac{1}{1-\beta^2} = \dfrac{-\beta^2}{1-\beta^2} = -\,\gamma^2\beta^2, \quad \beta = \dfrac{v}{c}$.

Substituting this expression for $1 - \gamma^2$ in equation (5.52), we obtain

$$\mathbf{E} = \frac{e\,\gamma\,\mathbf{r}}{r^3[\,\gamma^2 - \gamma^2\beta^2\sin^2\theta]^{3/2}}$$

$$= \frac{e\,(1-\beta^2)}{r^2[\,1-\beta^2\sin^2\theta]^{3/2}}\,\frac{\mathbf{r}}{r} \ . \qquad (5.53a)$$

That is, at $t = 0$ the electric field of a point charge moving rapidly with uniform velocity is still a radial field, varies inversely as the square of the distance r, but is no longer spherically symmetric.

The magnitude of the electric field at $\theta = 0^\circ$ or 180°, $E_{\theta=0,180}$, is given by

$$E_{\theta=0} = E_{\theta=180} = \frac{e}{r^2}(1-\beta^2) = \frac{e}{\gamma^2 r^2},$$

while in the transverse direction, the magnitude of the electric field is

$$E_{90} = \frac{e}{r^2}\frac{1}{\sqrt{1-\beta^2}} = \frac{e}{r^2}\gamma$$

so that $$\frac{E_{90}}{E_{0,180}} = \gamma^3, \quad \gamma > 1.$$

That is, in the transverse direction the electric field is γ^3 times as large as it is along the line of motion of the charge.

Similarly we can show that the magnetic field **H** can be written as

$$\mathbf{H} = \frac{e}{c}\frac{(1-\beta^2)\,\mathbf{v}\wedge\mathbf{r}}{r^3[\,1-\beta^2\sin^2\theta]^{3/2}}. \tag{5.53b}$$

Equation (5.53b) shows that the magnetic field **H** is perpendicular to the plane of **r** and **v**, and therefore, has no component in the direction of motion. Hence the lines of magnetic field are concentric circles around the direction of motion.

For $\frac{v}{c} << 1$, $\beta^2 = v^2/c^2$ is negligible as compared to 1. Therefore, for this case, equations (5.53a,b) reduce to

$$\mathbf{E} = \frac{e}{r^3}\mathbf{r} \tag{5.54a}$$

$$\mathbf{H} = \frac{e}{c}\frac{\mathbf{v}\wedge\mathbf{r}}{r^3}. \tag{5.54b}$$

Equation (5.54a) shows that for small velocities the electric field approximates to that of the stationary charge while equation (5.54b) represents the well known Biot-Savart law for a magnetic field.

Comparison of equations (5.53) and (5.54) shows that due to the uniform motion of the charge its field strength is decreased along its line of motion ($\theta = 0^\circ$ or 180°), while in the transverse direction ($\theta = 90^\circ$), the electric field is stronger than the Coulomb field (see Fig. 5.4) because of the extra factors $(1 - \beta^2)$.

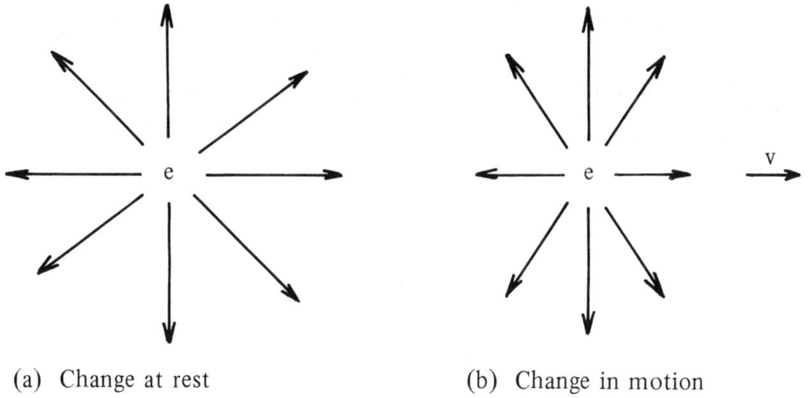

Fig. 5.4. Electric field produced by (a) a stationary charge, and (b) a moving charge

Notice that we have compared the classical and relativistic relations for the fields at the instant t = 0 when the moving charge is at the origin O of the frame S. However, at other instants the field will look the same, although translated to the right by an amount vt.

Potential due to an Arbitrarily Moving Point Charge

We will first calculate the potential at a point due to a uniformly moving point charge.

Suppose that a point charge e, moving with a uniform velocity **v**, is, at time t_1, at the origin O of a frame of reference S, whose x-axis is taken along the direction of motion of the charge, as shown in Fig. 5.5. The electromagnetic radiations starting from this charge e at time t_1 will reach a point P at a later time t so that

$$t - t_1 = \frac{r}{c},$$

where r is the distance OP and c is the velocity of electromagnetic radiations. By that time, the charge e, moving with uniform velocity v along the x-axis, would have reached another point, say O′. Thus at time t when the charge e is at O′, the electromagnetic field which had started from it when it was at O, has reached the point P. Hence when calculating the potential at the point P at time t, we have to calculate it due to the charge located at time t_1 = t −

r/c, and not at time t. For this reason this potential is called the **retarded potential** and (t − r/c) is called the **retarded time.** This retarded potential can be calculated easily by making use of the results of the special relativity.

Consider a frame of reference S′ fixed to the charge, having it at its origin O′, its x′-axis coincident with x-axis of S and its y′- and z′-axes parallel respectively to y- and z-axes of S. In this frame, due to this charge, the scalar and vector potentials at the point P(x′, y′, z′) at any time t′ are given by

$$\phi' = \frac{e}{r'}, \quad \mathbf{A}' = 0, \tag{5.55}$$

where $r' = O'P = \sqrt{x'^2 + y'^2 + z'^2}$.

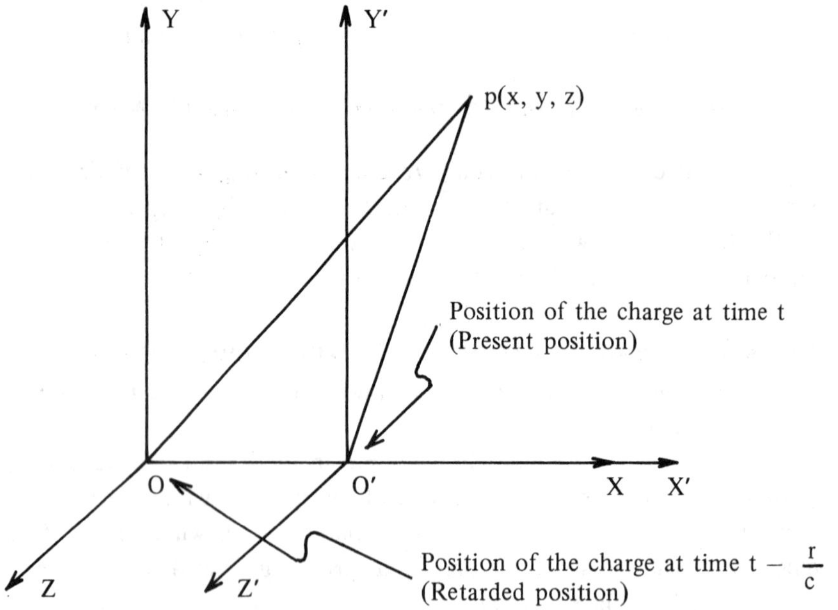

Fig. 5.5. Present and retarded positions of the charge in motion

Since the frame S′ is moving with the velocity **v** (as the moving charge is fixed in it) relative to S along their common axis, the potential at the point P(x, y, z) in S can be obtained from equations (5.55) by applying inverse transformation equations. Thus we have

$$A_x = \gamma\,(A_x' + \frac{v}{c}\phi') = \gamma\,\frac{v}{c}\frac{e}{r'} \qquad (5.56a)$$

$$A_y = A_y' = 0 \qquad (5.56b)$$

$$A_z = A_z' = 0 \qquad (5.56c)$$

$$\phi = \gamma\,(\phi' + \frac{v}{c}A_x') = \gamma\,\frac{e}{r'}. \qquad (5.56d)$$

However, by using the Lorentz transformation, we have

$$r' = \sqrt{x'^2 + y'^2 + z'^2} = \sqrt{\gamma^2\,(x - v\,t)^2 + y^2 + z^2}\,.$$

Substituting this expression for r' in equation (5.56d), we get

$$\phi = \frac{\gamma\,e}{\sqrt{\gamma^2\,(x - v\,t)^2 + y^2 + z^2}}. \qquad (5.57)$$

This gives us the scalar potential at the point P at a time t due to the charge e located at O at the time $t - r/c$.

Moreover:

$$A_x = \gamma\,\frac{v}{c}\frac{e}{r'} = \frac{v}{c}\,\frac{\gamma\,e}{\sqrt{\gamma^2\,(x - v\,t)^2 + y^2 + z^2}}$$

$$A_y = 0 = A_z,$$

so that $\mathbf{A} = \left(\frac{v}{c}\,\frac{\gamma\,e}{\sqrt{\gamma^2\,(x - v\,t)^2 + y^2 + z^2}}\,,\,0,\,0\right).$ (5.58)

This is the expression for the vector potential in S.

Now that we have obtained the scalar and vector potentials from a point charge moving at a constant velocity, we can calculate the electromagnetic field due to an arbitrarily moving charged particle (which may be moving with a speed nearly equal to that of light) in the following manner:

The electric field is given by

$$\mathbf{E} = -\,\mathrm{grad}\,\phi - \frac{1}{c}\frac{\partial \mathbf{A}}{\partial t}\,. \qquad (5.2b')$$

We can obtain the expression of grad ϕ and $\partial \mathbf{A}/\partial t$ by differentiating equation

(5.57) with respect to x, y, z and equation (5.58) with respect to t. First differentiating equation (5.57) with respect to x, we get

$$\frac{\partial \phi}{\partial x} = -\frac{\gamma^3 e (x - v t)}{[\gamma^2 (x - v t)^2 + y^2 + z^2]^{3/2}} .$$

Similarly, we obtain

$$\frac{\partial \phi}{\partial y} = -\frac{-\gamma e y}{[\gamma^2 (x - v t)^2 + y^2 + z^2]^{3/2}}$$

$$\frac{\partial \phi}{\partial z} = -\frac{-\gamma e z}{[\gamma^2 (x - v t)^2 + y^2 + z^2]^{3/2}}$$

so that

$$-\operatorname{grad} \phi = \frac{\gamma e}{[\gamma^2 (x - v t)^2 + y^2 + z^2]^{3/2}} (\gamma^2 (x - v t), y, z).$$

Next differentiating equation (5.58) with respect to t, we get

$$\frac{\partial \mathbf{A}}{\partial t} = (\frac{v^2}{c} \frac{\gamma^3 e (x - v t)}{[\gamma^2 (x - v t)^2 + y^2 + z^2]^{3/2}}, 0, 0).$$

Substituting these expressions for grad ϕ and $\partial \mathbf{A}/\partial t$ in equation (5.2b′), we get

$$\mathbf{E} = \frac{\gamma e}{[\gamma^2 (x - v t)^2 + y^2 + z^2]^{3/2}} (\gamma^2 (x - v t) - \frac{\gamma^2 v^2}{c^2} (x - v t), y, z)$$

or $$\mathbf{E} = \frac{\gamma e}{[\gamma^2 (x - v t)^2 + y^2 + z^2]^{3/2}} (x - v t, y, z). \qquad (5.59)$$

The magnetic field is obtained from the relation $\mathbf{H} = \operatorname{curl} \mathbf{A}$ by making use of relation (5.58):

$$\mathbf{H} = \begin{vmatrix} \frac{\partial}{\partial x} & \frac{\partial}{\partial y} & \frac{\partial}{\partial z} \\ A_x & A_y & A_z \\ \mathbf{i} & \mathbf{j} & \mathbf{k} \end{vmatrix}$$

$$= \left(0, \frac{\partial A_x}{\partial z}, -\frac{\partial A_x}{\partial y} \right)$$

$$= \frac{v}{c} \frac{\gamma e}{[\gamma^2 (x - v\,t)^2 + y^2 + z^2]^{3/2}} (0, -z, y). \tag{5.60}$$

To express the scalar and vector potentials in a form usually given in the literature, we proceed as follows:

From Fig. 5.5, we have

$$c = \frac{r'}{t'} = \frac{r}{t}, \tag{5.61}$$

so that $r' = c\,t' = c\,\gamma\,(t - \frac{v\,x}{c^2})$,

where we have made use of the Lorentz transformation.

Substituting the expression for t from equation (5.61) in the above relation, we obtain

$$r' = \gamma\,(r - \frac{v}{c} x)$$

or $$r' = \gamma\,(r - \frac{\mathbf{v} \cdot \mathbf{r}}{c}). \tag{5.62}$$

Then equation (5.56d) can be written as

$$\phi = \frac{\gamma e}{r'} = \frac{\gamma e}{\gamma\,(r - \frac{\mathbf{v} \cdot \mathbf{r}}{c})} = \frac{e}{(r - \frac{\mathbf{v} \cdot \mathbf{r}}{c})}. \tag{5.63}$$

To emphasize that the potential at P(x, y, z) at time t is to be calculated due to the charge in the position occupied by it at time $t - r/c$, we usually write the above relation as

$$\phi = \frac{e}{[r - \frac{\mathbf{v} \cdot \mathbf{r}}{c}]_{t' = t - r/c}} . \tag{5.64a}$$

Similarly, we can show that

$$\mathbf{A} = \frac{e}{c} \frac{\mathbf{v}}{[r - \frac{\mathbf{v} \cdot \mathbf{r}}{c}]_{t' = t - r/c}} . \tag{5.64b}$$

Problem

Derive relation (5.64b).

For a uniformly moving charge, **v** will have the same value at all times. Since the vector and scalar potentials depend only on the retarded velocity of the charge and not on its acceleration, equations (5.64) give the correct potentials even for an arbitrarily moving charge. They are known as the **Lienard-Wiechert potentials** of a moving charge.

Electromagnetic Field due to an Arbitrarily Moving Point Charge

We have seen that the scalar and vector potentials due to a point charge e moving with an arbitrary time-varying velocity v(t′) are given by

$$\phi = \frac{e}{\left[r - \frac{\mathbf{v}\cdot\mathbf{r}}{c}\right]_{t'=t-r/c}} \tag{5.64a$'$}$$

$$\mathbf{A} = \frac{e}{c}\frac{\mathbf{v}}{\left[r - \frac{\mathbf{v}\cdot\mathbf{r}}{c}\right]_{t'=t-r/c}} \quad . \tag{5.64b$'$}$$

Writing the vector **ß** for **v**/c and the unit vector **n** for **r**/r, we get

$$\phi = e/[r(1 - \frac{\mathbf{v}}{c}\cdot\frac{\mathbf{r}}{r})]_{ret} = e/[r(1 - \boldsymbol{\beta}\cdot\mathbf{n})]_{ret} \tag{5.65a}$$

$$\mathbf{A} = e\,[\,\frac{\mathbf{v}}{c}/\{r(1 - \boldsymbol{\beta}\cdot\mathbf{n})\}]_{ret} = e\,[\boldsymbol{\beta}/\{r(1 - \boldsymbol{\beta}\cdot\mathbf{n})\}]_{ret}, \tag{5.65b}$$

where for the sake of brevity we have written "ret" for the retarded potential indicating that a quantity in square brackets with "ret" as subscript is to be evaluated at the retarded time t′ = t − r(t′)/c. If we denote 1 − **ß**•**n** by k, we may write equations (5.65) as

$$\phi = e\left(\frac{1}{k\,r}\right)_{ret} \tag{5.66a}$$

$$\mathbf{A} = e\left[\frac{\boldsymbol{\beta}}{k\,r}\right]_{\text{ret}} . \tag{5.66b}$$

The electric and magnetic fields **E** and **H** can be determined from the above relations by making use of the formulae

$$\mathbf{E} = -\operatorname{grad}\phi - \frac{1}{c}\frac{\partial \mathbf{A}}{\partial t} \tag{5.2b$'$}$$

and

$$\mathbf{H} = \operatorname{curl}\mathbf{A}, \tag{5.2a$'$}$$

but this is a very tedious procedure. We will, therefore, first put ϕ and **A** in the integral form by making use of Dirac's delta function and then determine **E** and **H**.

We introduce the delta function $\delta\left(t' - t + \frac{r(t')}{c}\right)$ so that

$$\phi = e\int \frac{1}{r}\,\delta\left(t' - t + \frac{r(t')}{c}\right) dt' \tag{5.67}$$

$$\mathbf{A} = e\int \frac{\boldsymbol{\beta}}{r}\,\delta\left(t' - t + \frac{r(t')}{c}\right) dt' . \tag{5.68}$$

To show that equation (5.67) is equivalent to equation (5.66a), we define a variable u by the equation

$$u = t' - t + \frac{r(t')}{c} . \tag{5.69}$$

Then

$$\frac{du}{dt'} = 1 + \frac{1}{c}\frac{dr(t')}{dt'} . \tag{5.70}$$

Now we know that $r^2(t') = \mathbf{r}^2(t')$, so that, on differentiation with respect to t', we obtain

$$2\,r\frac{dr}{dt'} = 2\,\mathbf{r}\cdot\frac{d\mathbf{r}}{dt'}$$

or $$\frac{dr}{dt'} = \frac{\mathbf{r}}{r} \cdot \frac{d\mathbf{r}}{dt'} = \mathbf{n} \cdot (-\mathbf{v})$$

$$= -c\,\mathbf{n} \cdot \frac{\mathbf{v}}{c} = -c\,\mathbf{n} \cdot \boldsymbol{\beta}. \tag{5.71}$$

The negative sign in $d\mathbf{r}/dt' = -\mathbf{v}$ is due to the fact that the radius vector **r** has been drawn from the charge e to the observation point and not in the reverse order.

Substituting the above expression for dr/dt' in equation (5.70), we get

$$\frac{du}{dt'} = 1 + \frac{1}{c}(-c\,\boldsymbol{\beta} \cdot \mathbf{n}) = 1 - \boldsymbol{\beta} \cdot \mathbf{n} = k$$

or $$dt' = \frac{1}{k}\,du \tag{5.72}$$

Equation (5.67) therefore takes the form

$$\phi = e \int \frac{1}{r}\,\delta(u)\,\frac{du}{k}$$

$$= e\left(\frac{1}{k\,r}\right)_{u=0} = e\left(\frac{1}{k\,r}\right)_{t'=t-r/c} = e\left(\frac{1}{k\,r}\right)_{ret}$$

which is nothing but equation (5.66a).

Problem

Show that the expressions for **A** given by equations (5.68) and (5.66b) are equivalent.

After writing the scalar and vector potentials ϕ and **A** in the integral form as given by equations (5.67) and (5.68), we proceed further to calculate the electric and magnetic fields by using these equations.

Now $$\mathbf{E} = -\,\text{grad}\,\phi - \frac{1}{c}\frac{\partial \mathbf{A}}{\partial t}\,. \tag{5.2b'}$$

But, by virtue of equation (5.67):

$$\text{grad}\,\phi = \mathbf{n}\frac{\partial}{\partial r}\int \frac{e}{r}\,\delta(t' - t + \frac{r}{c})\,dt'$$

$$= e\int \mathbf{n}\frac{\partial}{\partial r}\,[\frac{1}{r}\,\delta(t' - t + \frac{r}{c})]\,dt'$$

$$= e\int \mathbf{n}\,[\,(-\frac{1}{r^2})\,\delta(t' - t + \frac{r}{c})$$

$$+ \frac{1}{r}\frac{\partial(t' - t + r/c)}{\partial r}\,\frac{\partial}{\partial(t' - t + r/c)}\,\delta(t' - t + \frac{r}{c})\,]\,dt'$$

$$= e\int [\,-\frac{\mathbf{n}}{r^2}\,\delta(t' - t + \frac{r}{c}) + \frac{\mathbf{n}}{r}\frac{1}{c}\,\delta'(t' - t + \frac{r}{c})]\,dt', \quad (5.73)$$

where the prime on the delta function indicates differentiation with respect to its own arguments

and
$$\frac{\partial \mathbf{A}}{\partial t} = \frac{\partial}{\partial t}\int \frac{e}{r}\,\boldsymbol{\beta}\,\delta(t' - t + \frac{r}{c})\,dt'$$

$$= \int \frac{e}{r}\,\boldsymbol{\beta}\,\frac{\partial}{\partial t}\,\delta(t' - t + \frac{r}{c})\,dt'$$

$$= \int \frac{e}{r}\,\boldsymbol{\beta}\,\frac{\partial(t' - t + r/c)}{\partial t}\,\frac{\partial}{\partial(t' - t + r/c)}\,\delta(t' - t + \frac{r}{c})\,dt'$$

$$= \int -\,\frac{e}{r}\,\boldsymbol{\beta}\,\delta'(t' - t + \frac{r}{c})\,dt'. \quad (5.74)$$

Substituting these expressions for grad ϕ and $\partial \mathbf{A}/\partial t$ in equation (5.2b′), we get

$$\mathbf{E} = e\int \Big[\frac{\mathbf{n}}{r^2}\delta(t' - t + \frac{r}{c}) - \frac{\mathbf{n}}{r}\frac{1}{c}\delta'(t' - t + \frac{r}{c})$$

$$+ \frac{1}{c}\frac{\mathbf{\beta}}{r}\delta'(t' - t + \frac{r}{c})\Big]\, dt'$$

or
$$\mathbf{E} = e\int \Big[\frac{\mathbf{n}}{r^2}\delta(t' - t + \frac{r}{c}) + \frac{\mathbf{\beta} - \mathbf{n}}{c\,r}\delta'(t' - t + \frac{r}{c})\Big]\, dt'. \tag{5.75}$$

Similarly, we can show that

$$\mathbf{H} = e\int \mathbf{n} \wedge \mathbf{\beta}\Big[-\frac{1}{r^2}\delta(t' - t + \frac{r}{c}) - \frac{1}{c\,r}\delta'(t' - t + \frac{r}{c})\Big]\, dt' \tag{5.76}$$

Problem

Derive relation (5.76).

To evaluate the integrals involving δ', we make use of the following property of the delta function:

$$\int f(x)\,\delta'(x - a)\,dx = -f'(a) \tag{5.77}$$

Then the first integral of equation (5.75) gives

$$\int \frac{\mathbf{n}}{r^2}\delta(t' - t + \frac{r}{c})\,dt' = \int \frac{\mathbf{n}}{r^2}\delta(u)\frac{du}{k} = \left(\frac{\mathbf{n}}{kr^2}\right)_{\text{ret}}.$$

The second integral on the right hand side of equation (5.75) yields

$$\int \frac{\mathbf{\beta} - \mathbf{n}}{c\,r}\delta'(t' - t + \frac{r}{c})\,dt' = \int \frac{\mathbf{\beta} - \mathbf{n}}{c\,r}\delta'(u)\frac{du}{k}$$

$$= -\Big[\frac{d}{du}\Big(\frac{\mathbf{\beta} - \mathbf{n}}{ckr}\Big)\Big]_{u=0} = -\Big[\frac{dt'}{du}\frac{d}{dt'}\Big(\frac{\mathbf{\beta} - \mathbf{n}}{ckr}\Big)\Big]_{t'=t-r/c}$$

$$= -\left[\frac{1}{k}\frac{d}{dt'}\left(\frac{\boldsymbol{\beta} - \mathbf{n}}{ckr}\right)\right]_{ret}.$$

Substituting these expressions in equation (5.75), we get

$$\mathbf{E} = e\left[\frac{\mathbf{n}}{k\,r^2} - \frac{1}{c\,k}\frac{d}{dt'}\left(\frac{\boldsymbol{\beta} - \mathbf{n}}{k\,r}\right)\right]_{ret}. \tag{5.78a}$$

Problem

Evaluate the integrals in equation (5.76) to show that

$$\mathbf{H} = e\left[\frac{\boldsymbol{\beta} \wedge \mathbf{n}}{k\,r^2} + \frac{1}{c\,k}\frac{d}{dt'}\left(\frac{\boldsymbol{\beta} - \mathbf{n}}{k\,r}\right)\right]_{ret}. \tag{5.78b}$$

We now intend to express the electric and magnetic fields in a form which does not involve the differentiation of any variable with respect to t′ except that of the vector **ß**. For this purpose, we shall derive certain relations which will be utilized later on.

We first prove that

$$\frac{d\mathbf{n}}{dt'} = \frac{c}{r}[\,\mathbf{n} \wedge (\,\mathbf{n} \wedge \boldsymbol{\beta}\,)\,] \tag{5.79}$$

We have earlier defined the unit vector **n** by the equation

$$\mathbf{n} = \frac{\mathbf{r}}{r}.$$

Differentiating with respect to t′, we get

$$\frac{d\mathbf{n}}{dt'} = \frac{d}{dt'}\left(\frac{\mathbf{r}}{r}\right) = \frac{1}{r}\frac{d\mathbf{r}}{dt'} + \mathbf{r}\frac{d}{dt'}\left(\frac{1}{r}\right)$$

$$= -\frac{1}{r}\mathbf{v} + \mathbf{r}\left(-\frac{1}{r^2}\right)\frac{dr}{dt'}, \qquad \text{because } \frac{d\mathbf{r}}{dt'} = -\mathbf{v}$$

$$= -\frac{c}{r}\frac{\mathbf{v}}{c} - \frac{1}{r}\frac{\mathbf{r}}{r}(\,-\,c\,\boldsymbol{\beta}\cdot\mathbf{n}\,),$$

where we have made use of equation (5.71).

or
$$\frac{d\mathbf{n}}{dt'} = -\frac{c}{r}\mathbf{ß} + \frac{c}{r}\mathbf{n}\,\mathbf{ß}\cdot\mathbf{n} = \frac{c}{r}(\mathbf{ß}\cdot\mathbf{n}\,\mathbf{n} - \mathbf{n}\cdot\mathbf{n}\,\mathbf{ß})$$
$$= \frac{c}{r}\mathbf{n}\wedge(\mathbf{n}\wedge\mathbf{ß}).$$

We next determine an expression for $\frac{d}{dt'}(k\,r)$. We have

$$\frac{d}{dt'}(k\,r) = \frac{d}{dt'}[(1-\mathbf{ß}\cdot\mathbf{n})\,r]$$
$$= \frac{dr}{dt'} - \frac{d}{dt'}(r\,\mathbf{ß}\cdot\mathbf{n})$$
$$= \frac{dr}{dt'} - \frac{dr}{dt'}\,\mathbf{ß}\cdot\mathbf{n} - r\,\mathbf{ß}\cdot\frac{d\mathbf{n}}{dt'} - \frac{d\mathbf{ß}}{dt'}\cdot\mathbf{n}$$
$$= -c\,\mathbf{ß}\cdot\mathbf{n} - (-c\,\mathbf{ß}\cdot\mathbf{n})\,\mathbf{ß}\cdot\mathbf{n} - c\,\mathbf{ß}\cdot[\mathbf{n}\wedge(\mathbf{n}\wedge\mathbf{ß})] - r\,\dot{\mathbf{ß}}\cdot\mathbf{n},$$

where we have written $\dot{\mathbf{ß}}$ for $d\mathbf{ß}/dt'$.

or
$$\frac{d}{dt'}(k\,r) = -c\,\mathbf{ß}\cdot\mathbf{n} + c\,(\mathbf{ß}\cdot\mathbf{n})^2 - c\,\mathbf{ß}\cdot(\mathbf{ß}\cdot\mathbf{n}\,\mathbf{n} - \mathbf{n}\cdot\mathbf{n}\,\mathbf{ß})$$
$$- r\,\dot{\mathbf{ß}}\cdot\mathbf{n}$$
$$= c\,\beta^2 - c\,\mathbf{ß}\cdot\mathbf{n} - r\,\dot{\mathbf{ß}}\cdot\mathbf{n}. \tag{5.80}$$

Now, from equation (5.78), performing the differentiation with respect to t′ in the second term on the right hand side of that equation, we get

$$\mathbf{E} = e\left[\frac{\mathbf{n}\,k}{k^2\,r^2} - \frac{1}{c\,k}\left\{\frac{d}{dt'}\left(\frac{\mathbf{ß}}{k\,r}\right) - \frac{d}{dt'}\left(\frac{\mathbf{n}}{k\,r}\right)\right\}\right]_{ret}$$
$$= e\left[\frac{\mathbf{n}}{k^2\,r^2}(1-\mathbf{n}\cdot\mathbf{ß}) - \frac{1}{c\,k}\frac{d}{dt'}\left(\frac{\mathbf{ß}}{k\,r}\right) + \frac{1}{c\,k^2\,r}\frac{d\mathbf{n}}{dt'}\right.$$
$$\left.+ \frac{\mathbf{n}}{c\,k}\frac{d}{dt'}\left(\frac{1}{k\,r}\right)\right]_{ret}. \tag{5.81}$$

But, by virtue of equation (5.79), we have

$$\frac{1}{c\,k^2\,r}\frac{d\mathbf{n}}{dt'} = \frac{1}{c\,k^2\,r}\frac{c}{r}\mathbf{n}\wedge(\mathbf{n}\wedge\mathbf{ß})$$

$$= e\Big[\frac{\mathbf{n}}{k^2 r^2}(1 - \mathbf{n}\cdot\boldsymbol{\beta}) - \frac{1}{c\,k}\frac{d}{dt'}\Big(\frac{\boldsymbol{\beta}}{k\,r}\Big) + \frac{1}{c\,k^2 r}\frac{d\mathbf{n}}{dt'}$$

$$+ \frac{\mathbf{n}}{c\,k}\frac{d}{dt'}\Big(\frac{1}{k\,r}\Big)\Big]_{ret}. \tag{5.81}$$

But, by virtue of equation (5.79), we have

$$\frac{1}{c\,k^2 r}\frac{d\mathbf{n}}{dt'} = \frac{1}{c\,k^2 r}\frac{c}{r}\mathbf{n}\wedge(\mathbf{n}\wedge\boldsymbol{\beta})$$

$$= \frac{1}{k^2 r^2}(\mathbf{n}\cdot\boldsymbol{\beta}\,\mathbf{n} - \boldsymbol{\beta}).$$

Substituting this expression in equation (5.81) and simplifying, we get

$$\mathbf{E} = e\Big[\frac{\mathbf{n}}{k^2 r^2} + \frac{\mathbf{n}}{c\,k}\frac{d}{dt'}\Big(\frac{1}{k\,r}\Big) - \frac{\boldsymbol{\beta}}{k^2 r^2} - \frac{1}{c\,k}\frac{d}{dt'}\Big(\frac{\boldsymbol{\beta}}{k\,r}\Big)\Big]_{ret}. \tag{5.82a}$$

Similarly, we can show that

$$\mathbf{H} = e\Big[\Big\{\frac{\boldsymbol{\beta}}{k^2 r^2} + \frac{1}{c\,k}\frac{d}{dt'}\Big(\frac{\boldsymbol{\beta}}{k\,r}\Big)\Big\}\wedge\mathbf{n}\Big]_{ret} \tag{5.82b}$$

Problem

Derive relation (5.82b).

It may be noticed from equations (5.82a) and (5.82b) that **E** and **H** are related by the equation

$$\mathbf{H} = \mathbf{n}\wedge\mathbf{E}, \tag{5.83}$$

i.e., the magnetic field **H** is everywhere perpendicular to the electric field **E** and to the unit vector **n** extending from the charge e at time t′ to the observation point.

Now $$\frac{d}{dt'}\left(\frac{1}{k\,r}\right) = -\frac{1}{k^2 r^2}\frac{d}{dt'}(k\,r) = -\frac{1}{k^2 r^2}\left(c\,\beta^2 - c\,\mathbf{n}\cdot\mathbf{ß} - r\,\mathbf{n}\cdot\dot{\mathbf{ß}}\right)$$

and $$\frac{d}{dt'}\left(\frac{\mathbf{ß}}{k\,r}\right) = \frac{1}{k\,r}\frac{d\mathbf{ß}}{dt'} + \mathbf{ß}\frac{d}{dt'}\left(\frac{1}{k\,r}\right)$$

$$= \frac{\dot{\mathbf{ß}}}{k\,r} - \frac{\mathbf{ß}}{k^2 r^2}\left(c\,\beta^2 - c\,\mathbf{n}\cdot\mathbf{ß} - r\,\mathbf{n}\cdot\dot{\mathbf{ß}}\right).$$

Substituting these expressions in equation (5.82a), we get

$$\mathbf{E} = e\Big[\frac{\mathbf{n}}{k^2 r^2} - \frac{\mathbf{n}}{c\,k^3 r^2}\left(c\,\beta^2 - c\,\mathbf{n}\cdot\mathbf{ß} - r\,\mathbf{n}\cdot\dot{\mathbf{ß}}\right) - \frac{\mathbf{ß}}{k^2 r^2}$$

$$- \frac{1}{c\,k^2 r}\dot{\mathbf{ß}} + \frac{\mathbf{ß}}{c\,k^3 r^2}\left(c\,\beta^2 - c\,\mathbf{n}\cdot\mathbf{ß} - r\,\mathbf{n}\cdot\dot{\mathbf{ß}}\Big]_{\text{ret}}. \qquad (5.84)$$

Writing $$\frac{\mathbf{n}}{k^2 r^2} = \frac{\mathbf{n}}{k^3 r^2}\,k = \frac{\mathbf{n}\,(1 - \mathbf{n}\cdot\mathbf{ß})}{k^3 r^2} = \frac{\mathbf{n}}{k^3 r^2} - \frac{\mathbf{n}\cdot\mathbf{ß}}{k^3 r^2}\mathbf{n}$$

and $$\frac{\mathbf{ß}}{k^2 r^2} = \frac{\mathbf{ß}}{k^3 r^2}\,k = \frac{\mathbf{ß}\,(1 - \mathbf{n}\cdot\mathbf{ß})}{k^3 r^2} = \frac{\mathbf{ß}}{k^3 r^2} - \frac{\mathbf{n}\cdot\mathbf{ß}}{k^3 r^2}\mathbf{ß},$$

and simplifying, we get

$$\mathbf{E} = e\Big[\frac{\mathbf{n}}{k^3 r^2} - \frac{\mathbf{ß}^2}{k^3 r^2}\,\mathbf{n} + \frac{1}{c\,k^3 r}\mathbf{n}\cdot\dot{\mathbf{ß}}\,\mathbf{n} - \frac{\mathbf{ß}}{k^3 r^2} - \frac{1}{c\,k^2 r}\dot{\mathbf{ß}}$$

$$+ \frac{\mathbf{ß}^2}{k^3 r^2}\,\mathbf{ß} - \frac{1}{c\,k^3 r}\mathbf{n}\cdot\dot{\mathbf{ß}}\;\mathbf{ß}\,\Big]_{\text{ret}}$$

$$= e\Big[\frac{1}{k^3 r^2}\{\mathbf{n} - \mathbf{ß} - \beta^2(\mathbf{n} - \mathbf{ß})\} + \frac{1}{c\,k^3 r}\{\,\mathbf{n}\cdot\dot{\mathbf{ß}}\,\mathbf{n} - k\,\dot{\mathbf{ß}}$$

$$- \mathbf{n} \cdot \dot{\boldsymbol{\beta}}\, \boldsymbol{\beta} \}]_{ret}. \tag{5.85}$$

The expression in braces can be written as

$$\mathbf{n} \cdot \dot{\boldsymbol{\beta}}\, \mathbf{n} - (1 - \mathbf{n} \cdot \boldsymbol{\beta})\, \dot{\boldsymbol{\beta}} - \mathbf{n} \cdot \dot{\boldsymbol{\beta}}\, \boldsymbol{\beta}$$

$$= \mathbf{n} \cdot \dot{\boldsymbol{\beta}}(\mathbf{n} - \boldsymbol{\beta}) - (\mathbf{n} \cdot \mathbf{n} - \mathbf{n} \cdot \boldsymbol{\beta})\, \dot{\boldsymbol{\beta}}, \text{ because } \mathbf{n} \cdot \mathbf{n} = 1.$$

$$= \mathbf{n} \cdot \dot{\boldsymbol{\beta}}\, (\mathbf{n} - \boldsymbol{\beta}) - \mathbf{n} \cdot (\mathbf{n} - \boldsymbol{\beta})\, \dot{\boldsymbol{\beta}}$$

$$= \mathbf{n} \wedge [(\mathbf{n} - \boldsymbol{\beta}) \wedge \dot{\boldsymbol{\beta}}].$$

Substituting this expression in equation (5.85) and simplifying, we get

$$\mathbf{E} = e \left[\frac{1}{k^3 r^2} (\mathbf{n} - \boldsymbol{\beta})(1 - \beta^2) \right]_{ret}$$

$$+ \frac{e}{c} \left[\frac{1}{k^3 r} (\mathbf{n} \wedge \{(\mathbf{n} - \boldsymbol{\beta}) \wedge \dot{\boldsymbol{\beta}}\}) \right]_{ret}. \tag{5.86a}$$

This expression does not involve any differentiation with respect to t' except that of the vector β.

The electric field thus consists of two parts. The first part does not depend upon the acceleration of the charged particle and varies at large distances like $1/r^2$. The second part depends upon the acceleration of the charged particle, and varies at large distances like $1/r$.

Problem

Show that the corresponding expression for the magnetic field **H** is given by

$$\mathbf{H} = \mathbf{n} \wedge \mathbf{E} = e \left[\frac{1}{k^3 r^2} \boldsymbol{\beta} \wedge \mathbf{n}\, (1 - \beta^2) \right]_{ret}$$

$$+ \frac{e}{c}\left[\frac{1}{k^3 r}\mathbf{n} \wedge (\mathbf{n} \wedge \{(\mathbf{n} - \mathbf{ß})\wedge\dot{\mathbf{ß}}\})\right]_{ret}. \qquad (5.86b)$$

For a charged particle moving with uniform velocity, $\dot{\mathbf{ß}} = 0$ and the expressions for **E** and **H** given by equation (5.86) become

$$\mathbf{E} = e\left[\frac{1}{k^3 r^2}(\mathbf{n} - \mathbf{ß})(1 - \beta^2)\right]_{ret} \qquad (5.87a)$$

$$\mathbf{H} = e\left[\frac{1}{k^3 r^2}\mathbf{ß} \wedge \mathbf{n}\,(1 - \beta^2)\right]_{ret}. \qquad (5.87b)$$

Problem

Show that the above expressions for **E** and **H** are equivalent to those given by equations (5.51).

The Electromagnetic Energy-Momentum Tensor

We have seen that the electric and magnetic fields are synthesized as an electromagnetic field tensor $f_{\mu\nu}$. We now define a new tensor $S_{\mu\sigma}$ of rank 2 in terms of this field tensor by the equation

$$S_{\mu\sigma} = \frac{1}{4\pi} f_{\mu\nu} f_{\sigma\nu} - \frac{1}{16\pi}\delta_{\mu\sigma} f_{\nu\gamma} f_{\nu\gamma}. \qquad (5.88)$$

The significance of this tensor stems from the fact that from its divergence emerge the laws of conservation of energy and momentum for an electromagnetic field. This tensor is called the energy-momentum tensor of the electromagnetic field. Notice that $S_{\mu\sigma} = S_{\sigma\mu}$, i.e., the energy-momentum tensor is symmetric. To deduce the laws of conservation of energy and momentum for the electromagnetic field, we proceed as follows. The divergence of $S_{\mu\sigma}$ is obtained by differentiating equation (5.88) with respect to x_σ:

$$\frac{\partial S_{\mu\sigma}}{\partial x_\sigma} = \frac{1}{4\pi}\frac{\partial f_{\mu\nu}}{\partial x_\sigma} f_{\sigma\nu} + \frac{1}{4\pi} f_{\mu\nu}\frac{\partial f_{\sigma\nu}}{\partial x_\sigma} - \frac{1}{8\pi}\delta_{\mu\sigma}\frac{\partial f_{\nu\gamma}}{\partial x_\sigma} f_{\nu\gamma}.$$

Since $\delta_{\mu\sigma} = 0$ or 1 according as σ is different from or equal to μ, this reduces to

$$\frac{\partial S_{\mu\sigma}}{\partial x_\sigma} = \frac{1}{4\pi}\frac{\partial f_{\mu\nu}}{\partial x_\sigma} f_{\sigma\nu} + \frac{1}{4\pi} f_{\mu\nu}\frac{\partial f_{\sigma\nu}}{\partial x_\sigma} - \frac{1}{8\pi}\frac{\partial f_{\nu\gamma}}{\partial x_\mu} f_{\nu\gamma}. \quad (5.89)$$

Now the expression $\frac{\partial f_{\mu\nu}}{\partial x_\sigma} f_{\sigma\nu}$ can be written as

$$\frac{\partial f_{\mu\nu}}{\partial x_\sigma} f_{\sigma\nu} = \frac{1}{2}\frac{\partial f_{\mu\nu}}{\partial x_\sigma} f_{\sigma\nu} + \frac{1}{2}\frac{\partial f_{\mu\nu}}{\partial x_\sigma} f_{\sigma\nu}.$$

Interchanging dummy indices ν and σ in the second term on the right hand side of this equation and making use of the antisymmetry property of the field tensor in the same term, we get

$$\frac{\partial f_{\mu\nu}}{\partial x_\sigma} f_{\sigma\nu} = \frac{1}{2}\left(\frac{\partial f_{\mu\nu}}{\partial x_\sigma} f_{\sigma\nu} + \frac{\partial f_{\mu\sigma}}{\partial x_\nu} f_{\nu\sigma}\right).$$

$$= \frac{1}{2}\left(\frac{\partial f_{\mu\nu}}{\partial x_\sigma} + \frac{\partial f_{\sigma\mu}}{\partial x_\nu}\right) f_{\sigma\nu}.$$

By using the second set of Maxwell's equations in the form given by equation (5.29), this reduces to

$$\frac{\partial f_{\mu\nu}}{\partial x_\sigma} f_{\sigma\nu} = \frac{1}{2}\frac{\partial f_{\sigma\nu}}{\partial x_\mu} f_{\sigma\nu}.$$

Writing ν and γ for the dummy indices σ and ν respectively on the right hand side of the above relation, we get

$$\frac{\partial f_{\mu\nu}}{\partial x_\sigma} f_{\sigma\nu} = \frac{1}{2}\frac{\partial f_{\nu\gamma}}{\partial x_\mu} f_{\nu\gamma}.$$

Substituting this expression in equation (5.89), we obtain

$$\frac{\partial S_{\mu\sigma}}{\partial x_\sigma} = \frac{1}{4\pi} f_{\mu\nu}\frac{\partial f_{\sigma\nu}}{\partial x_\sigma} = -\frac{1}{4\pi} f_{\mu\nu}\frac{\partial f_{\nu\sigma}}{\partial x_\sigma}. \quad (5.90)$$

Using the first set of Maxwell's equations, viz.,

$$\frac{\partial f_{\nu\sigma}}{\partial x_\sigma} = \frac{4\pi}{c} J_\nu, \quad (5.28')$$

equation (5.90) reduces to

$$\frac{\partial S_{\mu\sigma}}{\partial x_\sigma} = -\frac{1}{c} f_{\mu\nu} J_\nu.$$

Replacing the dummy index σ by ν, we get

$$\frac{\partial S_{\mu\nu}}{\partial x_\nu} = -\frac{1}{c} f_{\mu\nu} J_\nu. \tag{5.91}$$

Let us first consider the case when there is no source, i.e., $\rho = 0$. Then $J_\nu = (\mathbf{J}, i c \rho) = (\rho \mathbf{v}, i c \rho) = 0$ and equation (5.91) reduces to

$$\frac{\partial S_{\mu\nu}}{\partial x_\nu} = 0. \tag{5.92}$$

That is, in the absence of the sources, the divergence of the energy-momentum tensor is equal to zero.

For $\mu = 4$, this equation yields

$$\frac{\partial S_{4\nu}}{\partial x_\nu} = 0.$$

Summing over the dummy index ν, we obtain

$$\frac{\partial S_{41}}{\partial x_1} + \frac{\partial S_{42}}{\partial x_2} + \frac{\partial S_{43}}{\partial x_3} + \frac{\partial S_{44}}{\partial x_4} = 0. \tag{5.93}$$

But $$S_{\mu\sigma} = \frac{1}{4\pi} f_{\mu\nu} f_{\sigma\nu} - \frac{1}{16\pi} \delta_{\mu\sigma} f_{\nu\gamma} f_{\nu\gamma}. \tag{5.88$'$}$$

For $\mu = 4$, $\sigma = 1$, this gives

$$S_{41} = \frac{1}{4\pi} f_{4\nu} f_{1\nu} = \frac{1}{4\pi} (f_{41} f_{11} + f_{42} f_{12} + f_{43} f_{13} + f_{44} f_{14}).$$

Substituting the values of $f_{\mu\nu}$ from equation (5.21) in the above relation, we get

$$S_{41} = \frac{i}{4\pi} (E_2 H_3 - H_2 E_3) = \frac{i}{4\pi} (\mathbf{E} \wedge \mathbf{H})_1, \tag{5.94a}$$

where $(\mathbf{E} \wedge \mathbf{H})_j$ stands for the jth component of $\mathbf{E} \wedge \mathbf{H}$.
Similarly, we can show that

$$S_{42} = \frac{i}{4\pi} (\mathbf{E} \wedge \mathbf{H})_2 \tag{5.94b}$$

$$S_{43} = \frac{i}{4\pi} (\mathbf{E} \wedge \mathbf{H})_3. \tag{5.94c}$$

Again, from equation (5.88′), for $\mu = \sigma = 4$, we have

$$S_{44} = \frac{1}{4\pi} f_{4\nu} f_{4\nu} - \frac{1}{16\pi} f_{\nu\gamma} f_{\nu\gamma}. \tag{5.94d}$$

Now the expression $f_{4\nu} f_{4\nu}$ on the right hand side of this equation can be written as

$$f_{4\nu} f_{4\nu} = f_{41}^2 + f_{42}^2 + f_{43}^2 + f_{44}^2 = - E_1^2 - E_2^2 - E_3^2.$$

The product $f_{\nu\gamma} f_{\nu\gamma}$ in the second term can be expanded as

$$f_{\nu\gamma} f_{\nu\gamma} = f_{1\gamma} f_{1\gamma} + f_{2\gamma} f_{2\gamma} + f_{3\gamma} f_{3\gamma} + f_{4\gamma} f_{4\gamma}.$$

But $\quad f_{1\gamma} f_{1\gamma} = f_{11}^2 + f_{12}^2 + f_{13}^2 + f_{14}^2 = H_3^2 + H_2^2 - E_1^2$

$$f_{2\gamma} f_{2\gamma} = f_{21}^2 + f_{22}^2 + f_{23}^2 + f_{24}^2 = H_3^2 + H_1^2 - E_2^2$$

$$f_{3\gamma} f_{3\gamma} = H_2^2 + H_1^2 - E_3^2$$

$$f_{4\gamma} f_{4\gamma} = - E_1^2 - E_2^2 - E_3^2.$$

All these expressions have been obtained by using equation (5.21) for the field tensor $f_{\mu\nu}$. Substituting them in equation (5.94d) and simplifying, we get

$$S_{44} = - \frac{1}{16\pi} [\, 2\,(E_1^2 + E_2^2 + E_3^2) + 2\,(H_1^2 + H_2^2 + H_3^2)]$$

$$= - \frac{1}{8\pi} (E^2 + H^2) = - U, \tag{5.94e}$$

where $U = (E^2 + H^2)/8\pi$ is the volume energy density of the electromagnetic field.

Substituting the expressions for S_{41}, S_{42}, S_{43} and S_{44} in equation (5.93), we get

$$\frac{\partial}{\partial x_j} \frac{i}{4\pi}(\mathbf{E} \wedge \mathbf{H})_j + \frac{\partial}{\partial x_4} (-U) = 0$$

or $$\operatorname{div} (\frac{c}{4\pi} \mathbf{E} \wedge \mathbf{H}) + \frac{\partial U}{\partial t} = 0$$

or $$\operatorname{div} \mathbf{P} + \frac{\partial U}{\partial t} = 0, \tag{5.95}$$

where $$\mathbf{P} = \frac{c}{4\pi} \mathbf{E} \wedge \mathbf{H} \tag{5.96}$$

is called the **Poynting vector**. Equation (5.95) expresses the law of conservation of energy for the electromagnetic field in the differential form. The change of energy density per unit time is equal to the net inflow of the energy current per unit area.

To obtain the law of conservation of momentum, we again consider equation (5.92). For $\mu = 1$, we have

$$\frac{\partial S_{1\nu}}{\partial x_\nu} = 0.$$

Summing over the dummy index ν, we have

$$\frac{\partial S_{11}}{\partial x_1} + \frac{\partial S_{12}}{\partial x_2} + \frac{\partial S_{13}}{\partial x_3} + \frac{\partial S_{14}}{\partial x_4} = 0 \tag{5.93$'$}$$

or $$\operatorname{div} \mathbf{S}_1 + \frac{\partial S_{14}}{\partial x_4} = 0, \quad \text{where} \quad \mathbf{S}_1 = (S_{11}, S_{12}, S_{13}). \tag{5.97}$$

Now, from equation (5.94a), we have

$$S_{14} = S_{41} = \frac{i}{4\pi} (\mathbf{E} \wedge \mathbf{H})_1.$$

Therefore $$\frac{\partial S_{14}}{\partial x_4} = -\frac{i}{c} \frac{\partial S_{14}}{\partial t} = \frac{1}{4\pi c} \frac{\partial}{\partial t} (\mathbf{E} \wedge \mathbf{H})_1.$$

Substituting this expression in equation (5.97), we get

$$\operatorname{div} \mathbf{S}_1 + \frac{1}{c}\frac{\partial}{\partial t} \left[\frac{1}{4\pi} (\mathbf{E} \wedge \mathbf{H})_1\right] = 0. \tag{5.98}$$

However, the electromagnetic momentum density **G** is given by

$$\mathbf{G} = \frac{\mathbf{P}}{c^2}. \tag{5.99}$$

Then, by using equation (5.96), we have

$$G_1 = \frac{P_1}{c^2} = \frac{1}{4\pi c}(\mathbf{E} \wedge \mathbf{H})_1 \tag{5.100}$$

Substituting this expression for $\frac{1}{4\pi c}(\mathbf{E} \wedge \mathbf{H})_1$ in equation (5.98), we get

$$\text{div } \mathbf{S}_1 + \frac{\partial G_1}{\partial t} = 0. \tag{5.101a}$$

Similarly, we can show that

$$\text{div } \mathbf{S}_2 + \frac{\partial G_2}{\partial t} = 0. \tag{5.101b}$$

$$\text{div } \mathbf{S}_3 + \frac{\partial G_3}{\partial t} = 0. \tag{5.101c}$$

The last three equations can be put together as

$$\text{div } \mathbf{S}_a + \frac{\partial G_a}{\partial t} = 0, \tag{5.102}$$

where $a = 1, 2, 3$. These equations show that G_1, G_2, G_3 and consequently **G** are conserved. Hence equations (5.102) express the *law of conservation of momentum in the electromagnetic field in the differential form*. They are usually referred to as **Maxwell's stress relations.**

It may be noticed that the mere fact that equations (5.95) and (5.102) are equations of continuity for the energy density and the momentum density does not ensure that (**P**, U) or (S_a, G_a) form 4-vectors; in fact they do not. It should also be realized that the equation

$$\frac{\partial S_{\mu\nu}}{\partial x_\nu} = 0$$

which, in the absence of sources, leads to the laws of conservation of energy and momentum cannot be used to define the tensor $S_{\mu\nu}$ uniquely, because we can always add a divergence-free tensor to $S_{\mu\nu}$ without changing this equation.

We will now find an expression for the tensor S_{ij}, i, j = 1, 2, 3 in terms of the electric and magnetic fields. This tensor is called **Maxwell's stress**

tensor.

Writing i, j for μ, σ in equation (5.88), we get

$$S_{ij} = \frac{1}{4\pi} f_{i\nu} f_{j\nu} - \frac{1}{16\pi} \delta_{ij} f_{\nu\gamma} f_{\nu\gamma}.$$

Splitting the first term on the right hand side into its *space* and *time* parts, and substituting the expression for $f_{\nu\gamma} f_{\nu\gamma}$, which we have already determined, we get

$$S_{ij} = \frac{1}{4\pi} f_{ik} f_{jk} + \frac{1}{4\pi} f_{i4} f_{j4} - \frac{1}{16\pi} \delta_{ij} 2 (H^2 - E^2). \tag{5.103}$$

By making use of the expression for the field tensor $f_{\mu\nu}$, we obtain

$$f_{ik} f_{jk} = \begin{cases} H^2 - H_i^2 & \text{for } j = i \\ - H_i H_j & \text{for } j \neq i \end{cases} \qquad = \delta_{ij} H^2 - H_i H_j$$

$$f_{i4} f_{j4} = - E_i E_j.$$

Substituting these expressions in equation (5.103), we obtain

$$S_{ij} = \frac{1}{4\pi} (\delta_{ij} H^2 - H_i H_j - E_i E_j) - \frac{1}{8\pi} \delta_{ij} (H^2 - E^2)$$

$$= \frac{1}{8\pi} \delta_{ij} (H^2 + E^2) - \frac{1}{4\pi} (H_i H_j + E_i E_j). \tag{5.104}$$

This is the required expression for Maxwell's stress tensor S_{ij} in terms of the electric and magnetic fields.

PROBLEMS

1. A particle of charge e moves in an electromagnetic field with velocity **v**. Verify that the quantities

$$e\gamma\left(\mathbf{E} + \frac{1}{c}\mathbf{v} \wedge \mathbf{H}\right), \quad e\gamma\,\frac{1}{c^2}\mathbf{v}\cdot\mathbf{E}$$

transform like **r**, t.

2. Uniform constant electric and magnetic fields **E** and **H** are parallel to the axes OY and OZ respectively of a rectangular Cartesian frame O(XYZ). A charged particle is moving in the half-plane OXY ($Y>0$). Use the relativistic equations of motion to determine possible paths when $E < H$.

(*London University*)

6

RELATIVISTIC KINEMATICS OF BINARY COLLISIONS

Introduction

Collision theory has played an important role in the development of high energy physics, as it can be used to write scattering amplitudes for a particular reaction, which, in turn, determine the differential and total cross sections. These cross sections can be measured experimentally and thus used to test the validity of a particular model. This sheds light on the nature of particles involved in the reaction. For a process involving two incoming and two outgoing particles, one or more scattering amplitudes completely describe the reaction, the number of amplitudes being determined by the spin content of the reaction. Each of these amplitudes depends only on two kinematical variables, viz., the total energy and the scattering angle. But the values of these variables in a reaction depend upon the choice of the frame of reference; they are different in different frames of reference. For instance, the total energy of a system in the CM frame will have a value different from that in the laboratory frame. In the relativistic treatment of scattering, it is therefore convenient to express the scattering amplitudes in terms of those variables which are Lorentz invariant, i.e., which have the same value in all inertial frames of reference and therefore are not affected by the choice of a particular inertial frame. We will first try to find these Lorentz invariant variables.

Mandelstam Variables

To determine such kinematic invariants, consider a process of the type

$$a_1 + a_2 \rightarrow a_3 + a_4$$

which is customarily symbolized as shown in Fig. 6.1. Let m_i be the rest mass, $\mathbf{p}_i$ the 3-momentum, E_i the energy and p_i the 4-momentum of the particle a_i, $i = 1, 2, 3, 4$. Since the square of the magnitude of a 4-vector as well as the scalar product of any two 4-vectors is an invariant quantity, we can construct

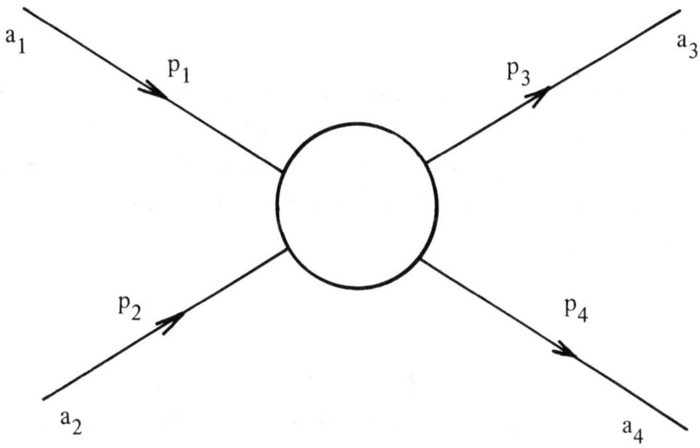

Fig.6.1. Diagram for two-body process

the following ten invariants (and their functions) from the 4-momenta p_1, p_2, p_3, p_4:

$$p_1^2,\ p_2^2,\ p_3^2,\ p_4^2,\ p_1\,p_2,\ p_1\,p_3,\ p_1\,p_4,\ p_2\,p_3,\ p_2\,p_4,\ p_3\,p_4\ .$$

Problem

Show that the scalar product of any two 4-vectors is an invariant quantity.

We shall use the natural system of units in which $c = \hbar = 1$ so that for the ith particle, the energy, mass and 3-momentum are related by the equation

$$E_i^2 = m_i^2 + \mathbf{p}_i^2 \tag{6.1}$$

Now the square of the magnitude of the 4-vector $p_i = (\mathbf{p}_i, iE_I)$ is given by

$$p_i^2 = (\mathbf{p}_i^2 - E_i^2) \tag{6.2}$$

Eliminating E_i^2 between equations (6.1) and (6.2), we get

$$p_i^2 = - m_i^2, \tag{6.3}$$

which is said to be the **mass shell condition** for a particle of mass m_i. Therefore the first four invariants, being equal to $- m_1^2$, $- m_2^2$, $- m_3^2$, $- m_4^2$, respectively, are all constants and hence uninteresting. The remaining six invariants are not all independent. They are related by the 4-momentum conservation equation

$$p_1 + p_2 = p_3 + p_4, \tag{6.4}$$

which yields four relations. The four conditions contained in these equations reduce the number of independent invariants to two. However, we should be careful to select only those two invariants which are independent of each other. It is customary to select $-(p_1 + p_2)^2$ and $-(p_1 - p_3)^2$ as the two independent invariants and call them s and t respectively:

$$s = -(p_1 + p_2)^2 \tag{6.5}$$

$$t = -(p_1 - p_3)^2. \tag{6.6}$$

The variables s and t, being negatives of the squares of 4-vectors, are relativistically invariant.

For the sake of symmetry and for convenience in the computations for backward scattering, it is usual to define another variable u, which is *not* independent of s and t, by the equation

$$u = -(p_1 - p_4)^2. \tag{6.7}$$

The variables s, t, u are called **Mandelstam variables** and are also denoted by s_3, s_2, s_1 respectively. These three variables are linearly dependent and are such that

$$\begin{aligned} s + t + u &= -(p_1 + p_2)^2 - (p_1 - p_3)^2 - (p_1 - p_4)^2 \\ &= -p_1^2 - p_2^2 - 2p_1 p_2 - p_1^2 - p_3^2 \\ &\qquad + 2 p_1 p_3 - p_1^2 - p_4^2 + 2 p_1 p_4 \end{aligned}$$

$$= - p_1^2 - p_2^2 - p_3^2 - p_4^2 - 2p_1 (p_1 + p_2 - p_3 - p_4)$$

$$= m_1^2 + m_2^2 + m_3^2 + m_4^2,$$

where we have made use of equàtions (6.3) and (6.4),

or $$s + t + u = \Sigma\, m_i^2 . \qquad (6.8)$$

For an elastic collision

$$a_1 + a_2 \rightarrow a_1 + a_2 ,$$

we have $m_1 = m_3 = m$, say, and $m_2 = m_4 = \mu$, say, so that equation (6.8) yields

$$s + t + u = 2\, m^2 + 2\, \mu^2. \qquad (6.9)$$

If the particles in the elastic collision are of the same mass, we have $m_1 = m_2 = m_3 = m_4 = m$, say. Equation (6.8) then reduces to

$$s + t + u = 4\, m^2. \qquad (6.10)$$

Physical Significance of s

Let us first see what the physical significance of s is. We have

$$s = - (p_1 + p_2)^2 = - (\mathbf{p}_1 + \mathbf{p}_2, i(E_1 + E_2))^2.$$

But in the CM frame, 3-momentum of the system is zero. Denoting the total energy of the system in the CM frame by ε, we get

$$s = - (0, i\varepsilon)^2 = \varepsilon^2. \qquad (6.11)$$

This shows that s is the square of the CM energy of the system. Note that this expression for s is true for all types of collisions.

Momentum, Energy and Scattering Angle in the CM Frame in Terms of Mandelstam Variables

Let us next consider the reaction $a + b \rightarrow c + d$, also written as

$a_1 + a_2 \rightarrow a_3 + a_4$ or $1 + 2 \rightarrow 3 + 4$ or $ab \rightarrow cd$, and express the magnitudes of 3-momenta and energies of the particles involved in the reaction and the scattering angle in the CM frame in terms of Mandelstam variables. We will now denote the various quantities as shown below:

Reaction	1 +	2 →	3 +	4
Rest Masses	m_1	m_2	m_3	m_4
Energies	E_1	E_2	E_3	E_4
4-Momenta	p_1	p_2	p_3	p_4
3-Momenta	$\mathbf{p}_1$	$\mathbf{p}_2$	$\mathbf{p}_3$	$\mathbf{p}_4$
Magnitudes of 3-Momenta	k_1	k_2	k_3	k_4

It may be noted that in the CM frame, by definition, the momentum of the system is zero, so that 3-momenta of the two particles before as well as after the collision are equal and opposite: $\mathbf{p}_2 = -\mathbf{p}_1$, $\mathbf{p}_4 = -\mathbf{p}_3$ and $k_2 = k_1 = k$, say, while $k_4 = k_3 = k'$.

We will first express the magnitudes k and k' of the 3-momenta of colliding and emerging particles respectively in the CM frame in terms of the Mandelstam variable s. For particles 1 and 2, we have

$$E_1 = \sqrt{k^2 + m_1^2}$$

and $$E_2 = \sqrt{k^2 + m_2^2}$$

so that $$(E_1 + E_2)^2 = \left(\sqrt{k^2 + m_1^2} + \sqrt{k^2 + m_2^2}\right)^2$$

or $$\varepsilon^2 = k^2 + m_1^2 + k^2 + m_2^2 + 2\left(\sqrt{k^2 + m_1^2}\right)\left(\sqrt{k^2 + m_2^2}\right),$$

where $\varepsilon = E_1 + E_2$ and represents total energy of the system before collision.

or $$(\varepsilon^2 - m_1^2 - m_2^2 - 2k^2)^2 = 4\,(k^2 + m_1^2)(k^2 + m_2^2).$$

Simplifying, we get

$$(\varepsilon^2 - m_1^2 - m_2^2)^2 - 4k^2 \varepsilon^2 = 4 m_1^2 m_2^2$$

or
$$k = \frac{\sqrt{(\varepsilon^2 - m_1^2 - m_2^2)^2 - 4 m_1^2 m_2^2}}{2 \varepsilon}$$

or
$$k = \frac{\sqrt{(\varepsilon^2 - m_1^2 - m_2^2)^2 - 4 m_1^2 m_2^2}}{2 \sqrt{s}}, \tag{6.12}$$

where $s = \varepsilon^2$ is the total energy of the system.

We now define a quantity $\lambda(x, y, z)$ by the equation

$$\lambda(x, y, z) = \sqrt{(x - y - z)^2 - 4 y z} . \tag{6.13}$$

Then equation (6.12) may be written as

$$k = \frac{\lambda(s, m_1^2, m_2^2)}{2 \sqrt{s}}. \tag{6.14}$$

Similarly, we can show that

$$k' = \frac{\lambda(s, m_3^2, m_4^2)}{2 \sqrt{s}}. \tag{6.15}$$

Equations (6.14) and (6.15) express the magnitudes of 3-momenta, k and k′, in terms of the invariant variable s.

Problem

Show that

$$k^2 = \frac{1}{4 s} [s - (m_1 + m_2)^2] [s - (m_1 - m_2)^2].$$

Problem

Show that

$$s = 2k^2 + m_1^2 + m_2^2 + 2 \sqrt{(k^2 + m_1^2) (k^2 + m_2^2)} \cdot$$

To express the energies in terms of the Mandelstam variables, we proceed as follows. We know that

$$E_1 = \sqrt{k^2 + m_1^2}$$

but
$$k^2 + m_1^2 = \frac{1}{4\,s}\left[(s - m_1^2 - m_2^2)^2 - 4\,m_1^2\,m_2^2\right] + m_1^2$$
$$= \frac{1}{4\,s}\left[(s - m_2^2)^2 + m_1^4 - 2m_1^2\,s + 2\,m_1^2\,m_2^2\right.$$
$$\left. - 4\,m_1^2\,m_2^2 + 4m_1^2\,s\right]$$
$$= \frac{1}{4\,s}\left[(s - m_2^2)^2 + m_1^4 + 2m_1^2\,s - 2\,m_1^2\,m_2^2\,\right]$$
$$= \frac{(s - m_2^2 + m_1^2)^2}{4\,s}$$

so that $\sqrt{k^2 + m_1^2} = \dfrac{(s + m_1^2 - m_2^2)^2}{2\sqrt{s}}$.

This shows that
$$E_1 = \frac{(s + m_1^2 - m_2^2)^2}{2\sqrt{s}} \qquad (6.16a)$$

Similarly, we can prove that

$$E_2 = \frac{(s + m_2^2 - m_1^2\,)^2}{2\sqrt{s}} \qquad (6.16b)$$

$$E_3 = \frac{(s + m_3^2 - m_4^2\,)^2}{2\sqrt{s}} \qquad (6.16c)$$

and
$$E_4 = \frac{(s + m_4^2 - m_3^2\,)^2}{2\sqrt{s}}\ . \qquad (6.16d)$$

So far, we have expressed the energies and the magnitudes of 3-momenta of the particles in the reaction $1 + 2 \rightarrow 3 + 4$ in terms of the Lorentz invariant variable s. Let us now see how we can express the scattering angle θ for the deflection from 1 to 3 in terms of Mandelstam variables s, t and u. We have

$$t = -\,(p_1 - p_3)^2 = -\,(\mathbf{p}_1 - \mathbf{p}_3,\, i(E_1 - E_3))^2$$

$$= -\,(\mathbf{p}_1^2 + \mathbf{p}_3^2 - 2\mathbf{p}_1 \cdot \mathbf{p}_3 - (E_1 - E_3)^2\,)$$

$$= -\,(\mathbf{p}_1^2 + \mathbf{p}_3^2 - 2k\,k'\,\cos\theta - E_1^2 - E_3^2 + 2E_1\,E_3)$$

$$= m_1^2 + m_3^2 + 2k\,k'\,\cos\theta - 2E_1\,E_3, \tag{6.17}$$

where we have used relation (6.1),

or
$$t = m_1^2 + m_3^2$$
$$+ \frac{1}{2\,s}\sqrt{(s-m_1^2-m_2^2)^2-4m_1^2m_2^2}\,\sqrt{(s-m_3^2-m_4^2)^2-4m_3^2m_4^2}\,\cos\theta$$
$$- \frac{1}{2\,s}(s + m_1^2 - m_2^2)\,(s + m_3^2 - m_4^2)$$

or
$$\cos\theta = \frac{2\,s\,(t - m_1^2 - m_3^2) + (s + m_1^2 - m_2^2)\,(s + m_3^2 - m_4^2)}{\lambda(s, m_1^2, m_2^2)\,\lambda(s, m_3^2, m_4^2)}\,. \tag{6.18}$$

This equation expresses the scattering angle θ in the CM frame in terms of the Mandelstam variables s and t.

The numerator on the right hand side of equation (6.18) can be written as

$$2\,s\,t - 2\,s\,m_1^2 - 2\,s\,m_3^2 + s^2 + s\,m_3^2 - s\,m_4^2 + s\,m_1^2 - s\,m_2^2$$
$$+ (m_1^2 - m_2^2\,)\,(m_3^2 - m_4^2\,)$$

$$= 2\,s\,t + s^2 - s\,(m_1^2 + m_2^2 + m_3^2 + m_4^2\,) + (m_1^2 - m_2^2\,)\,(m_3^2 - m_4^2\,)$$

$$= 2\,s\,t + s^2 - s\,(s + t + u) + (m_1^2 - m_2^2\,)\,(m_3^2 - m_4^2\,)$$

$$= 2\,s\,t + s^2 - s^2 - s\,t - s\,u + (m_1^2 - m_2^2\,)\,(m_3^2 - m_4^2\,)$$

$$= s\,(t - u) + (m_1^2 - m_2^2\,)\,(m_3^2 - m_4^2\,),$$

where we have made use of the relation

$$s + t + u = \Sigma\, m_i^2\,. \tag{6.8'}$$

Therefore equation (6.18) may be put in the form

$$\cos\theta = \frac{s\,(t-u) + (m_1^2 - m_2^2)\,(m_3^2 - m_4^2)}{\lambda(s, m_1^2, m_2^2)\,\lambda(s, m_3^2, m_4^2)}. \qquad (6.19a)$$

Problem

Show that

$$\sin\theta = 2\sqrt{s}\,[stu - s\,(m_1^2 - m_3^2)\,(m_2^2 - m_4^2)$$

$$- t\,(m_1^2 - m_2^2)\,(m_3^2 - m_4^2)$$

$$- (m_1^2 m_4^2 - m_2^2 m_3^2)\,(m_1^2 + m_4^2 - m_2^2 - m_3^2)]^{1/2}$$

$$/[\lambda\,(s, m_1^2, m_2^2)\,\lambda(s, m_3^2, m_4^2)]. \qquad (6.19b)$$

For an elastic collision, $a + b \to a + b$, the masses are related by the equations $m_1 = m_3$ and $m_2 = m_4$ so that equations (6.19) give

$$\cos\theta = \frac{s\,(t-u) + (m_1^2 - m_2^2)^2}{\lambda^2(s, m_1^2, m_2^2)} \qquad (6.20a)$$

$$\sin\theta = \frac{2\sqrt{s\,[s\,t\,u - t\,(m_1^2 - m_2^2)^2]^{1/2}}}{\lambda^2(s, m_1^2, m_2^2)}. \qquad (6.20b)$$

If the particles are of the same mass, $m_1 = m_2 = m$, say, then we have

$$\cos\theta = \frac{s\,(t-u)}{\lambda^2(s, m^2, m^2)} \qquad (6.21a)$$

$$\sin\theta = \frac{2\,s\sqrt{t\,u}}{\lambda^2(s, m^2, m^2)} \qquad (6.21b)$$

Problem

By making explicit calculations, show that

$$\left(\frac{s\,(t-u)}{\lambda^2(s, m^2, m^2)}\right)^2 + \left(\frac{2\,s\sqrt{t\,u}}{\lambda^2(s, m^2, m^2)}\right)^2 = 1.$$

In the CM frame, the Mandelstam variables s, t and u take a particularly simple form for an elastic collision between particles of equal masses. We have

$$s = -(p_1 + p_2)^2 = -(\mathbf{p}_1 + \mathbf{p}_2, i(E_1 + E_2))^2$$

But for an elastic collision between particles of equal masses, the energies of the particles are also equal: $E_1 = E_2 = E$, say. The above equation therefore reduces to

$$s = 4E^2 = 4(k^2 + m^2), \text{ because } E^2 = k^2 + m^2. \quad (6.22)$$

Moreover, we have

$$t = -(p_1 - p_3)^2 = -(\mathbf{p}_1 - \mathbf{p}_3, i(E_1 - E_3))^2.$$

Since for an elastic collision between particles of equal masses, $E_1 = E_3$ and $|\mathbf{p}_1| = |\mathbf{p}_3| = k$, the above equation yields

$$t = -(\mathbf{p}_1 - \mathbf{p}_3, 0)^2 = -(\mathbf{p}_1 - \mathbf{p}_3)^2 = -\Delta, \quad (6.23)$$

where Δ is the square of the 3-momentum transfer.

or
$$t = -\Delta = -\mathbf{p}_1^2 - \mathbf{p}_3^2 + 2\mathbf{p}_1 \cdot \mathbf{p}_3$$

$$= -k^2 - k^2 - 2k^2 \cos\theta$$

$$= -2k^2(1 - \cos\theta). \quad (6.24)$$

Problem

Show that, in the CM frame, for an elastic collision between particles of equal masses, the variable u is given by

$$u = -2k^2(1 + \cos\theta). \quad (6.25)$$

Physical Region in the Rectangular st-Plane

By using equations (6.22), (6.24) and (6.25), we shall describe the constraints imposed on the ranges of the variables s, t, u for an elastic process between particles of equal masses: $m_1 = m_2 = m_3 = m_4 = m$. Since k^2, the square of the magnitude of the 3-momentum of each of the colliding particles in the CM frame, must be real and non-negative, equation (6.22) yields

$$s \geq 4\,m^2. \tag{6.26a}$$

Moreover, as $\cos\theta$ can vary from -1 to $+1$, by virtue of equation(6.24) t can vary from $-4k^2$ to 0 so that

$$-\,4k^2 \leq t \leq 0. \tag{6.26b}$$

Similarly, we can deduce from equation (6.25) that

$$-\,4k^2 \leq u \leq 0. \tag{6.26c}$$

For a given process, the allowed values of these variables are said to determine the physical region for that process. Since only two of the variables are independent, the physical region can be depicted on a 2-dimensional plot with two of the three variables as coordinates.

We will now plot the physical region for an elastic collision between particles of equal masses in the st-plane. We have shown that for a physical region

$$s \geq 4m^2$$

$$t \leq 0$$

$$u \leq 0.$$

By virtue of the equation $s + t + u = 4m^2$, the last of these relations yields

$$(4m^2 - s) \leq t.$$

or

$$t \geq (4m^2 - s). \tag{6.27}$$

We are now well equipped to locate the physical region in the st-plane:

(i) Since $t \leq 0$, no part of the physical region lies above the s-axis
(ii) Since $s \geq 4m^2$, the physical region lies on and on the right of $s = 4m^2$. Combining (i) and (ii), we notice that the physical region lies in the fourth quadrant of the rectangular st-plane.
(iii) Moreover, for the physical region, we have

$$-t \leq (s - 4m^2).$$

The relation $t = -s + 4m^2$ is the equation of a straight line with intercept equal to $4m^2$ and slope equal to -1 so that it makes an angle of 135° with the

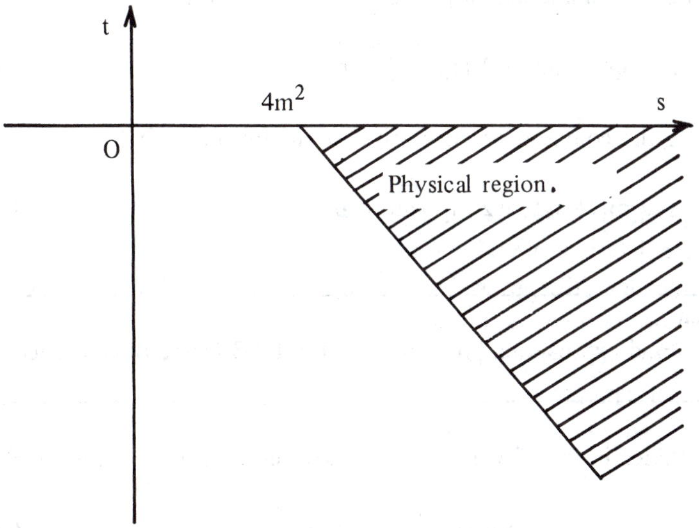

Fig.6.2. Physical region for the elastic collision between particles of equal masses

s-axis. This line is drawn in Fig. 6.2. The physical region lies in the fourth quadrant between the s-axis and the line $-t = s - 4m^2$ and is indicated by the shaded area in Fig. 6.2.

Expression for s in LAB frame

Let us next find the expression for s in the LAB frame of reference. Denoting the 3-momenta and energies of the incident particle and the target by $\mathbf{q}_1$, $\mathbf{q}_2$, w_1, w_2 respectively, we have

$$s = -(p_1 + p_2)^2 = -(\mathbf{q}_1 + \mathbf{q}_2, i(w_1 + w_2))^2 \tag{6.28}$$

In the laboratory, the target is stationary so that $\mathbf{q}_2 = 0$ and w_2 = rest mass energy of the stationary particle = m_2. Equation (6.28), therefore, may be written as

$$s = -(\mathbf{q}_1, i(w_1 + w_2))^2$$

$$= -(q_1^2 - w_1^2 - m_2^2 - 2\, m_2\, w_1)$$

$$= m_1^2 + m_2^2 + 2\, m_2\, w_1, \tag{6.29}$$

where we have made use of the relation $w_1^2 = \mathbf{q}_1^2 + m_1^2$.

or $$s = m_1^2 + m_2^2 + 2\, m_2 \sqrt{m_1^2 + q_1^2}\,. \tag{6.30}$$

Following the common practice, we write $\mathbf{p}_L$ for $\mathbf{q}_1$, so that

$$s = m_1^2 + m_2^2 + 2\, m_2 \sqrt{m_1^2 + p_L^2}\,. \tag{6.31}$$

Problem

Find expressions for t and u in the LAB frame of reference.

Since in the CM frame, $s = \varepsilon^2$, we may write equation (6.29), as

$$\varepsilon^2 = m_1^2 + m_2^2 + 2\, m_2\, w_1. \tag{6.32}$$

This equation gives a relation between the energy of the system in the CM frame and the energy of the incident particle in the LAB frame.

Relation between the Magnitudes of 3-Momenta of the Incident Particle in the LAB and CM Frames

Let us next see how the magnitude of the 3-momentum of the incident particle in the LAB frame is related to the magnitude of the 3-momentum of the same particle in the CM frame.

Let p_L and k be the magnitudes of 3-momenta of the incident particle in the laboratory frame and the CM frame respectively. Rearranging equation (6.31) and squaring terms, we obtain

$$(s - m_1^2 - m_2^2)^2 = 4\, m_2^2\, (m_1^2 + p_L^2)$$

or $$(s - m_1^2 - m_2^2)^2 - 4\, m_1^2\, m_2^2 = 4\, m_2^2\, p_L^2$$

or $$\lambda^2(s, m_1^2 - m_2^2) = 4\, m_2^2\, p_L^2$$

or $$\lambda(s, m_1^2 - m_2^2) = 2\, m_2\, p_L\,.$$

Substituting the expression for λ from equation (6.14) in the above equation, we get

$$k\sqrt{s} = m_2\, p_L$$

or $$\frac{k}{p_L} = \frac{m_2}{\sqrt{s}} = \frac{m_2}{\varepsilon} \qquad (6.33)$$

Thus if p_L and ε are given, k can be calculated.

Crossed Reactions

Consider the reaction

$$a_1 + a_2 \rightarrow a_3 + a_4, \qquad ((1))$$

$$p_1 \quad p_2 \quad p_3 \quad p_4,$$

where below each particle we have indicated its 4-momentum. This process is symbolized by Fig.6.1. Let us replace a_2 by $\bar{a}_3$ and a_3 by $\bar{a}_2$, where a bar indicates an antiparticle. Then we obtain a new process

$$a_1 + \bar{a}_3 \rightarrow \bar{a}_2 + a_4, \qquad ((2))$$

$$p_1 \quad -p_3 \quad -p_2 \quad p_4,$$

where a_1 and $\bar{a}_3$ are now the incoming particles while $\bar{a}_2$ and a_4 are the outgoing particles. This reaction is symbolized by Fig. 6.3. The 4-momenta of $\bar{a}_2$ and $\bar{a}_3$ are $-p_2$ and $-p_3$ respectively because the 3-momentum of an antiparticle has a sign opposite to that of a particle. This process is said to be a crossed process with respect to the given process ((1)), and is *assumed* to be possible in nature. In fact, from a given process $a_1 + a_2 \rightarrow a_3 + a_4$, we can obtain five reactions by all possible combinations of a_1, a_2, a_3 and a_4, an incoming (outgoing) particle of 4-momentum p being treated as an outgoing (incoming) antiparticle of 4-momentum −p. All these reactions are called **crossed reactions** and it is *assumed* that they will always be possible in nature. We write down all the crossed reactions which can be obtained from the given reaction

$$a_1 + a_2 \rightarrow a_3 + a_4. \qquad ((1))$$

These are

$$a_1 + \bar{a}_3 \rightarrow \bar{a}_2 + a_4 \qquad ((2))$$

$$a_1 + \bar{a}_4 \rightarrow a_3 + \bar{a}_2 \qquad ((3))$$

$$\bar{a}_3 + a_2 \rightarrow \bar{a}_1 + a_4 \qquad ((4))$$

$$\bar{a}_3 + \bar{a}_4 \rightarrow \bar{a}_1 + \bar{a}_2 \qquad ((5))$$

$$\bar{a}_4 + a_2 \rightarrow a_3 + \bar{a}_1. \qquad ((6))$$

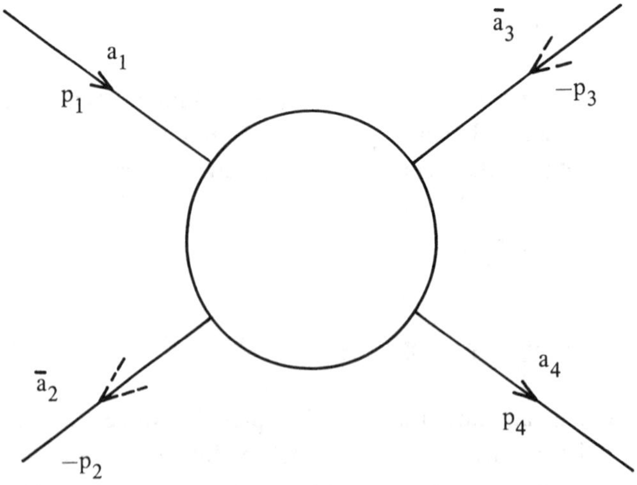

Fig. 6.3 Diagram for the process $a_1 + \bar{a}_3 \rightarrow \bar{a}_2 + a_4$

By making use of the invariance under time-reversal six more possible reactions can be obtained:

$$a_3 + a_4 \to a_1 + a_2 \qquad ((7))$$

$$\bar{a}_2 + a_4 \to a_1 + \bar{a}_3 \qquad ((8))$$

$$a_3 + \bar{a}_2 \to a_1 + \bar{a}_4 \qquad ((9))$$

$$\bar{a}_1 + a_4 \to \bar{a}_3 + a_2 \qquad ((10))$$

$$\bar{a}_1 + \bar{a}_2 \to \bar{a}_3 + \bar{a}_4 \qquad ((11))$$

$$a_3 + \bar{a}_1 \to \bar{a}_4 + a_2. \qquad ((12))$$

Example

We will write down all possible reactions from the given reaction

$$K^- n \to \pi^- \Lambda$$

which are

$$K^- \pi^+ \to \bar{n} \Lambda$$

$$K^- \bar{\Lambda} \to \pi^- \bar{n}$$

$$\pi^+ n \to K^+ \Lambda$$

$$\bar{\Lambda} n \to \pi^- K^+$$

$$\pi^+ \bar{\Lambda} \to K^+ \bar{n}$$

Six more reactions can be obtained by using invariance with respect to time-reversal.

---

s, t and u Channels

Let us again consider the reaction

$$a_1 + a_2 \rightarrow a_3 + a_4 \,.$$

$$p_1 \quad p_2 \quad p_3 \quad p_4$$

We have defined the kinematic invariants s, t, u by the equations

$$s = -(p_1 + p_2)^2$$

$$t = -(p_1 - p_3)^2$$

$$u = -(p_1 - p_4)^2$$

and have shown that s stands for the square of the CM energy of the system.

Let us next consider the crossed reaction

$$a_1 + \bar{a}_3 \rightarrow \bar{a}_2 + a_4 \,.$$

$$p_1 \quad -p_3 \quad -p_2 \quad p_4$$

If in this reaction we denote the Mandelstam variables by $\bar{s}, \bar{t}, \bar{u}$, then

$$\bar{s} = -(p_1 + p_3)^2$$

$$\bar{t} = -(p_1 - p_2)^2$$

$$\bar{u} = -(p_1 - p_4)^2 \,.$$

We note that

$$\bar{s} = -(p_1 - p_3)^2 = -(\mathbf{p}_1 - \mathbf{p}_3, i(E_1 - E_3))^2 \,.$$

But in the CM frame of reaction ((2)): $\mathbf{p}_1 - \mathbf{p}_3 = 0$ and $E_1 - E_3 =$ total CM energy of the $a_1\bar{a}_3$ system $= \varepsilon$, so that $s = -(0, i\varepsilon)^2 = \varepsilon^2$, i.e., $\bar{s}$ is the square of the CM energy for this reaction.

It may be noticed from the equations defining the Mandelstam variables for reactions((1)) and ((2)) that these variables are related by

$$s = \bar{t}, \; t = \bar{s}, \; u = \bar{u},$$

i.e., the two Mandelstam variables s and t interchange their role for reactions

((1)) and ((2)). For reaction ((1)), s represents the CM energy while for reaction ((2)), t or $\bar{s}$ represents the CM energy. For this reason any given reaction ((1)) for which s is always the CM energy is said to be an s-channel reaction while the crossed section ((2)) for which t is the CM energy is said to be the corresponding t-channel reaction.

Similarly, for the crossed reaction

$$a_1 + \bar{a}_4 \rightarrow a_3 + \bar{a}_2,$$

$$p_1 \quad -p_4 \quad p_3 \quad -p_2$$

we have

$$\hat{s} = -(p_1 - p_4)^2$$

$$\hat{t} = -(p_1 - p_3)^2$$

$$\hat{u} = -(p_1 + p_4)^2$$

so that $\quad s = \hat{u}\ ,\ t = \hat{t}\ ,\ u = \hat{s}\ ,$

i.e., u represents the square of the CM energy for this reaction which is therefore called a u-channel reaction.

Note that when the given process is an elastic process ab → ab, the corresponding u-channel also represents an elastic process ab → ab.

Example

For the reaction

$$\pi^- + p \rightarrow \pi^- + p, \qquad \text{(s-channel)}$$

the corresponding t- and u-channel reactions are given by

$$\pi^- + \pi^+ \rightarrow \bar{p} + p, \qquad \text{(t-channel)}$$

$$\pi^- + \bar{p} \rightarrow \pi^- + \bar{p}, \qquad \text{(u-channel)}$$

Physical Regions for s, t, u Channels

Let us now plot the physical regions for s, t, u channels for an elastic collision between particles of the same mass. Since only two of the Mandelstam variables are independent, all possible values of s, t, u can be represented on a 2-dimensional plot. This can be done conveniently by using trilinear coordinates. Let us first define these coordinates. Consider any triangle ABC and let P be any point in the plane of this triangle, as shown in Fig. 6.4. Let α, β, γ be the perpendicular distances of this point from the sides BC, CA and AB of this triangle so that these sides serve as coordinate axes with equations $\alpha = 0, \beta = 0$ and $\gamma = 0$, respectively. We shall take α, β, γ to be positive when directed towards any point lying inside the triangle. The directed distances α, β, γ are called the triangular or trilinear coordinates of the point P referred to the triangle ABC. It is known from geometry that the

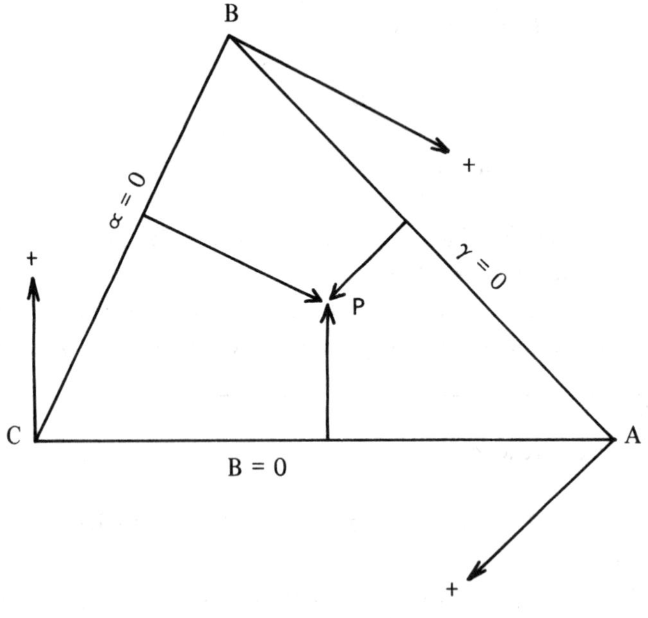

Fig. 6.4 Trilinear coordinates

distances α, β, γ are not all independent; their sum is a constant.

We are now well equipped to plot the physical region for an elastic collision between particles of the same mass in a plane with s, t, u as triangular coordinates of any point. The symmetry among s, t, u suggests the use of an equilateral triangle ABC as the reference triangle. Let us choose the sides BC, CA and AB of this triangle as the coordinate axes s = 0, t = 0, u = 0 respectively. We can satisfy the constraint $s + t + u = 4\ m^2$ by choosing the altitude of the triangle as equal to $4m^2$. Then the equation of the line passing through the point A and parallel to the line s = 0 is $s = 4\ m^2$. Moreover,

(i) as $t \leq 0$, no part of the physical region lies "above" the line t = 0,

(ii) since $s \geq 4\ m^2$, the physical region lies on and to the "right" of this line.

(iii) since $u \leq 0$, no part of the physical region lies "above" the line u = 0.

This s-channel physical region is indicated as shaded area I in Fig. 6.5.

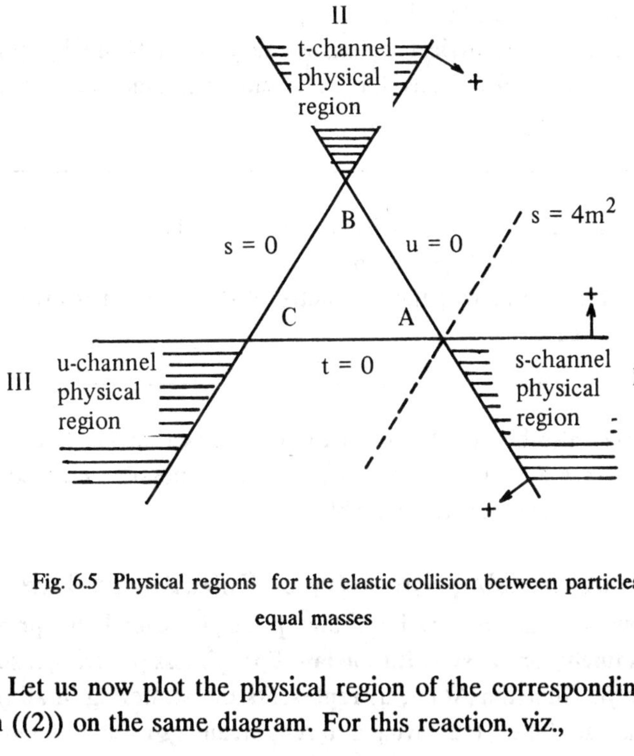

Fig. 6.5 Physical regions for the elastic collision between particles of equal masses

Let us now plot the physical region of the corresponding t-channel reaction ((2)) on the same diagram. For this reaction, viz.,

$$a_1 + \bar{a}_3 \rightarrow \bar{a}_2 + a_4, \qquad ((2))$$

$$p_1 \quad -p_3 \quad -p_2 \quad p_4$$

the Mandelstam variables are given by

$$\bar{s} = -(p_1 - p_3)^2 \geq 4m^2$$

$$\bar{t} = -(p_1 + p_2)^2 \leq 0$$

$$\bar{u} = -(p_1 - p_4)^2 \leq 0.$$

Therefore for the t-channel process, $s = \bar{t} \leq 0$, $t = \bar{s} \geq 4m^2$, $u = \bar{u} \leq 0$. Thusthe variables s and t have interchanged their roles while u has remained unaffected. The physical region for the t-channel will then be the shaded area II as shown in Fig. 6.5.

Similarly, we can plot the u-channel physical region. This is indicated as the shaded area III in Fig. 6.5.

Such a plot of physical regions is said to be a Mandelstam diagram. Note that the physical regions for s-, t- and u-channel reactions do not overlap.

Problem

Plot the physical region for the s-, t-, u-channel reactions for the elastic process $\pi^- p \to \pi^- p$.

Remark: Notice that the symmetry of the diagram is broken in this case.

It may also be noted that for the given reaction $a_1 + a_2 \to a_3 + a_4$, the decay process $a_1 \to \bar{a}_2 + a_3 + a_4$ will also be a possible crossed reaction provided that it is energetically possible.

Principle of Crossing Symmetry or Substitution Law

Now we state a very important principle, called the **principle of crossing symmetry** or the **substitution law.** For spinless particles, it states that a single analytic function A(s, t, u) represents the scattering in all the three channels, the variables s, t, u having different meanings in each channel. More specifically, if for a reaction involving spinless particles, we know the analytic expression for the s-channel scattering amplitudes A(s, t, u), then the t-channel scattering amplitude is given by the same analytic expression, the Lorentz invariant variables having different meanings in each channel. Since the physical regions for the three channels are disjointed, the connection between them is made by analytic continuation.

It may also be noted that as s, t, u are linearly dependent, the function A(s, t, u) is actually a function of any two of the three variables s, t, u. We may, therefore, write it as A(s, t) or A(t, u) or A(u, s).

Cross Sections in Terms of Mandelstam Variables

Suppose that f is the s-channel scattering amplitude for a reaction in the CM frame. Then we know that the differential cross section is given by the relation

$$\frac{d\sigma}{d\Omega} = \frac{k}{k'} f^*f = \frac{k}{k'} |f|^2. \qquad (6.34)$$

For an elastic collision, a + b → a + b, we have $k' = k$, and the above formula reduces to

$$\frac{d\sigma_{el}}{d\Omega} = | f_{el} |^2 . \qquad (6.35)$$

However, it is often convenient if we specify the differential cross section in terms of $d\sigma/d|t|$ instead of $d\sigma/d\Omega$. We will therefore find an expression for $d\sigma/d|t|$. We know that

$$\Omega = 2\pi(1 - \cos\theta)$$

so that

$$d\Omega = -2\pi \, d(\cos\theta). \qquad (6.36)$$

Moreover, we have proved that

$$\cos\theta = \frac{t - m_1^2 - m_3^2 + 2E_1 E_3}{2kk'}$$

so that

$$d(\cos\theta) = \frac{dt}{2kk'}.$$

Substituting this expression for $d(\cos\theta)$ in equation (6.36), we get

$$d\Omega = -\frac{\pi}{kk'} dt$$

or

$$d\Omega = \frac{\pi}{kk'} d(-t).$$

Since, for the physical region, $-t$ will always be non-negative, we may write the above equation as

$$d\Omega = \frac{\pi}{k\,k'}\, d|t|$$

or

$$\frac{d\Omega}{d|t|} = \frac{\pi}{k\,k'}. \tag{6.37}$$

Therefore,

$$\frac{d\sigma}{d|t|} = \frac{d\sigma\, d\Omega}{d\Omega\, d|t|} = \frac{k'}{k}|f|^2 \frac{\pi}{k\,k'} = \frac{\pi}{k^2}|f|^2,$$

where we have substituted the expression for $d\sigma/d\Omega$ and $(d\Omega/d|t|)$ from equations (6.34) and (6.37). It is customary to write t for $|t|$ so that the formula $\pi/k^2\,|f|^2$ may be written as

$$\frac{d\sigma}{dt} = \frac{\pi}{k^2}|f|^2. \tag{6.38}$$

Notice that now the formula for the differential cross section is the same for elastic as well as inelastic reactions.

We next define the field theoretic amplitude T by the equation

$$T = 8\pi f \sqrt{s} \tag{6.39}$$

so that, by substituting the expression for f from equation (6.39) in equation (6.38), the formula for the differential cross section becomes

$$\frac{d\sigma}{dt} = \frac{1}{64\pi k^2 s}|T|^2. \tag{6.40}$$

The total cross section may be obtained from the optical theorem according to which

$$\sigma_{tot} = \frac{4\pi}{k}\,\mathrm{Im}\, f_{el}(\cos\theta = 1, \varepsilon). \tag{6.41}$$

Writing $\dfrac{T_{el}(s, t=0)}{8\pi\sqrt{s}} = f_{el}$ and simplifying, we get

$$\sigma_{tot} = \frac{1}{2k\sqrt{s}}\,\mathrm{Im}T_{el}(s,t=0). \tag{6.42}$$

Let us see what form formulae (6.40) and (6.42) take at very high energies. We know that

$$k = \frac{\lambda(s, m_1^2, m_2^2)}{2\sqrt{s}} . \tag{6.14'}$$

Therefore, we may write

$$k^2 = \frac{(s - m_1^2 - m_2^2)^2 - 4\, m_1^2\, m_2^2}{4\, s} .$$

For very high energy of the system, the masses and their squares can be neglected as compared with s. Then the relation yields

$$k^2 \approx \frac{s^2}{4\, s}$$

or $$k \approx \frac{\sqrt{s}}{2} .$$

Substituting this expression for k in equations (6.40) and (6.42), we obtain

$$\frac{d\sigma}{dt} \approx \frac{1}{16\,\pi\, s^2} \left| T \right|^2 \tag{6.43}$$

$$\sigma_{tot} \approx \frac{1}{s} \operatorname{Im} T_{el}\,(s, t=0). \tag{6.44}$$

PROBLEMS

1. Show that

$$E_1 = m_2 w_1 + \frac{m_1^2}{\varepsilon}.$$

2. Consider the elastic reaction

$$\pi^- + p \rightarrow \pi^- + p,$$

produced by a beam of negative π-mesons incident on protons which are supposed to be at rest in the laboratory. Suppose that in the LAB frame the momentum of the incident pions is 12.5 GeV/c. Calculate the momentum of the pions in the CM frame.
Hint: Use the formulae:

$$s = m_1^2 + m_2^2 + 2\, m_2 \sqrt{m_1^2 + p_L^2} \quad \text{and} \quad k = \frac{m_2\, p_L}{\sqrt{s}}.$$

3. Show that the scattering angle $\phi \equiv \phi_{13}$ in the LAB frame in terms of the Mandelstam variables is given by

$$\cos\phi = [(s-m_1^2-m_2^2)\,(s-m_1^2-m_4^2) + t(s-m_2^2-m_1^2) - 2m_2^2(m_1^2+m_3^2)]$$

$$/\lambda(s, m_1^2, m_2^2)\, \lambda(m_2^2, m_1^2, m_3^2).$$

7

TACHYONS

According to the special theory of relativity, the mass and energy of a particle, moving with a velocity **v**, are given by the equations

$$m = \frac{m_0}{\sqrt{1 - v^2/c^2}} \tag{7.1a}$$

$$E = m\,c^2, \tag{7.1b}$$

where m_0 is the rest mass or the proper mass of the particle. If the speed of the particle is increased, its mass and hence energy would be enhanced in accordance with equations (7.1). However, these equations show that the particle cannot be accelerated to a velocity equal to that of light because then its mass and energy will become infinite. This does not mean that we cannot have particles moving with a velocity c. Photons and neutrinos have a velocity equal to that of light. But they are not accelerated from a lower speed to attain this velocity; they possess this velocity at the time of their creation. If they have to have finite mass and energy, they must possess zero rest mass. Thus the special relativity only recognizes the impossibility of accelerating a particle to the speed of light; it does not forbid a particle to have the speed of light at the time of its creation. Conversely, a particle moving with the velocity of light cannot be slowed down: the only way to slow it down is to make it disappear.

Let us now examine the question whether a particle can have a speed greater than that of light. We have already shown that the light barrier cannot be reached by a material particle. Therefore the question of crossing the light barrier by accelerating a particle does not arise. Let us therefore examine the possibility of the emission of a particle with a velocity u greater than that of light at the time of its creation during some as yet undiscovered process. For $u > c$, the equation $m = m_0/(1 - u^2/c^2)^{1/2}$ tells us that the mass of such a particle has to be imaginary. Bilaniuk *et al.*[4] have pointed out that this difficulty about the existence of particles moving with velocities greater than c can be overcome by assuming that the *proper mass of any such particle is pure imaginary*: $m_0 = i\,m_0^*$, where m_0^* is real. Then the mass of such a particle moving with a velocity $u > c$ is given by

$$m = \frac{m_0^*}{\sqrt{u^2/c^2 - 1}} \tag{7.2}$$

and is therefore a real number. m_0^* is called the **meta-mass.** The imaginary proper mass of a particle moving with a velocity $u > c$ does not create any problem because it is not an observable physical quantity. We, therefore, conclude that particles moving faster than light can exist. Such particles are called **tachyons.** This name was first suggested in 1967 by Feinberg and is derived from the Greek word **tachys,** meaning swift.

The energy and momentum of a tachyon will also be real, because

$$E = m\,c^2 = \frac{m_0^* c^2}{\sqrt{u^2/c^2 - 1}} \tag{7.3a}$$

$$\mathbf{p} = m\,\mathbf{u} = \frac{m_0^* \mathbf{u}}{\sqrt{u^2/c^2 - 1}} \tag{7.3b}$$

$$þ = \frac{m_0^* u}{\sqrt{u^2/c^2 - 1}}, \tag{7.3c}$$

where $þ = |\mathbf{p}|$.

The first of these equations shows that as u decreases the denominator will decrease and the energy of a tachyon will increase. This means that in order to decrease the speed of a tachyon, energy has to be supplied and not taken away from it. For $u \to c$, the energy $E \to \infty$, i.e., to decrease the speed of a tachyon to that of light, an infinite amount of energy has to be supplied to the tachyon. This is impossible; so that a tachyon would always be moving with a velocity greater than that of light. Thus approaching from higher velocities, the light barrier cannot be crossed or even reached by the tachyons.

It may be stressed that it has not been claimed that a particle with real proper mass can be accelerated to superluminal velocities; certainly not. The special theory of relativity, however, does not object to the existence of particles having velocities greater than c at the time of their creation.

Equations (7.3) show that the energy E and the magnitude of momentum þ decrease as the speed of the particle increases. For $u \to \infty$, the energy of the particle tends to zero. To evaluate þ as $u \to \infty$, we notice that as in this case unity can be ignored as compared to u^2/c^2, we may write

$$þ \approx \frac{m_0^* u}{u/c} = m_0^* c, \tag{7.4}$$

i.e., þ tends to a constant value $m_0^* c$.

Tachyons, if they exist, create a number of problems. We shall consider one of these problems and seek its solution. Suppose that, in a frame of reference S, a tachyon emitted from a source P and moving with a velocity $u>c$, is absorbed a while later by a sink Q; the source and the sink being at rest in this frame. Let Δx and Δt be the spatial and temporal intervals between the two events at P and Q as measured in this frame. Let $\Delta x'$ and $\Delta t'$ be the corresponding intervals between these two events as measured in a frame S′ moving with a velocity v relative to S. Then

$$\Delta t' = \gamma(v)\left(\Delta t - \frac{v}{c^2}\Delta x\right). \tag{7.5}$$

Since $u = \Delta x/\Delta t$, equation (7.5) may be written as

$$\Delta t' = \gamma(v)\,\Delta t\left(1 - \frac{uv}{c^2}\right) = \gamma(v)\Delta t\left(1 - \frac{v}{c^2/u}\right). \tag{7.6}$$

The factor c^2/u is always less than c; it is therefore possible to choose $v>c^2/u$ so that $\Delta t'$ is negative. In other words, there exists an infinite number of frames for which $\Delta t'$ is negative, i.e., in which the events occur in the reverse order; the event at P takes place after the event at Q. If this result were considered independently, it would mean that it is possible to choose a frame of reference in which a tachyon would strike its target even before it is emitted by its source. This violates the principle of causality which states that every cause precedes its effect. However, we shall show that, by postulating a new principle, this violation of the causality principle can be avoided provided this result is considered together with the concept of negative energy.

We know that the energy E′ of a particle as observed in S′ is related to its energy E and momentum **p** in the frame S by the equation

$$E' = \gamma(v)\,(E - v\,pc\,). \tag{7.7}$$

For a tachyon, $þ = uE/c^2$, so that the above equation may be written as

$$E' = \gamma(v)\,E\left(1 - \frac{uv}{c^2}\right). \tag{7.8}$$

Dividing equation (7.6) by equation (7.8), we get

$$\Delta t'/E' = \Delta t/E$$

or
$$E' = \frac{E}{\Delta t}\Delta t'. \tag{7.9}$$

Since E and Δt are positive, equation (7.9) shows that whenever $\Delta t' < 0$, the energy E would also be negative.

The difficulties of reversal of the time sequence and negative energy were overcome by Bilaniuk and Sudarshan[5] by postulating the "reinterpretation principle" according to which a negative-energy tachyon that has been absorbed first and emitted later is equivalent to a positive-energy tachyon that has been emitted first and absorbed later. Thus, in the frame S′, the negative-energy tachyon absorbed by Q before it is emitted by the source P is equivalent to a positive energy tachyon absorbed by Q after it had been emitted from the source P.

Experimentalists have been making all possible efforts in the search for tachyons since the idea was first floated by Bilaniuk *et al.*[4] in 1962. Even in vacuum, the tachyons will be travelling faster than light. Therefore, if they are charged, they would be emitting the Cerenkov radiation. Moreover, since in vacuum the tachyons are the only particles which can move faster than light, their identification would be unambiguous. In 1968, Alväger and Kreisler[6] attempted to detect tachyons, if produced, by bombarding lead with gamma rays from a 5 milli-curie Cesium-134 source, by exploiting Cerenkov radiation in vacuum. No trace of the tachyons was found. In 1977, Perepelitsa[7] also looked for the Cerenkov radiation emitted by charged tachyons in vacuum, but all in vain.

There is another class of experiments which have been performed for the detection of tachyons but which depend upon the interaction of tachyons with matter. If tachyons were produced in cosmic ray interaction in the upper atmosphere, they would reach the earth before the shower. Ramana Murthy[8] looked for such events before the arrival of the main shower at the earth. No such interaction was observed. Clay and Crouch[9] repeated this experiment with an improved experimental technique and detected what might be interpreted as tachyons. However, they have concluded that the non-random events which they detected prior to the extensive air showers might have been the result of fission or spallation in the interstellar medium. The result, therefore, remains inconclusive. The existence of tachyons is thus not proved.

PROBLEM

Professor X has invented a death-ray whose velocity is $\lambda c (\lambda > 1)$, and the discovery does not affect the usual relativity theory. He fires the ray when he is at rest in an inertial frame S but it is intercepted by a secret agent who is at rest in another inertial frame S′. If this agent can receive and transmit the ray with the same velocity prove that, so long as the velocity of S′ relative to S exceeds $2\lambda c/1+\lambda^2$, he can return the ray to reach Professor X **before** it has been transmitted. Assume next that the agent can receive the ray, but transmit only normal light signals. Prove that such a signal sent after receipt of the ray cannot then reach Professor X before he transmits the ray.

(*London University*)

APPENDIX A

THE MICHELSON-MORLEY EXPERIMENT

The Michelson-Morley experiment, which was performed in 1887 to determine the absolute velocity of the earth by performing an optical experiment confined to the earth, has played a significant role in the advent of the special theory of relativity. In this experiment, a monochromatic beam of light from a source S, made parallel by passing through a lens L, is split up

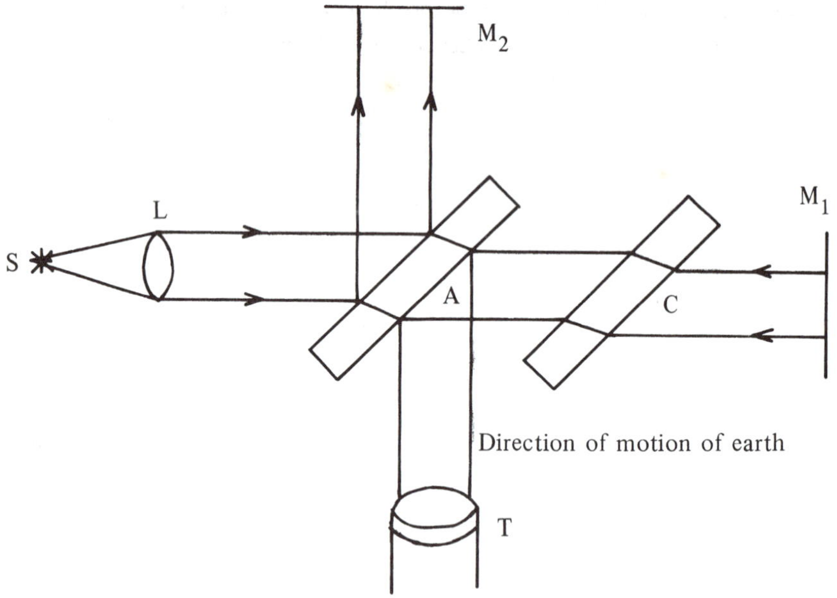

Fig. A-1. Schematic diagram of the Michelson-Morley experiment

into two parts by a thinly silvered glass mirror A inclined at an angle of 45° to the beam as shown in Fig. A.1. The two parts of the beam fall normally on the mirrors M_1 and M_2, which are fully silvered on their front surfaces, and are reflected back. They finally enter the telescope T as shown in Fig. A.1 and combine to produce an interference pattern in the telescope. The glass plate C has the same thickness and inclination as the plate A so that the two interfering light beams traverse the same distance through glass. To avoid strain on rotation, the whole apparatus is mounted on a heavy stone which floats in mercury.

If we assume that the velocity v of the earth through the ether is parallel to the light path from A to M_1 and c is the velocity of light, then the times taken by the light beams in the round trips parallel and perpendicular to the ether drift are, respectively, given by

$$t_1 = \frac{2 c \ell}{c^2 - v^2} \tag{A-1a}$$

$$t_2 = \frac{2 \ell}{\sqrt{c^2 - v^2}} \tag{A-1b}$$

Problem

Derive equations (A-1).

The difference Δt in times taken by the two beams is given by

$$\Delta t = t_1 - t_2 = \frac{2 c \ell}{c^2 - v^2} - \frac{2 \ell}{\sqrt{c^2 - v^2}}$$

$$= \frac{2 \ell}{c} \left(1 - \frac{v^2}{c^2}\right)^{-1} - \frac{2 \ell}{c} \left(1 - \frac{v^2}{c^2}\right)^{-1/2}$$

$$\approx \frac{2 \ell}{c} \left(1 + \frac{v^2}{c^2}\right) - \frac{2 \ell}{c} \left(1 + \frac{1}{2} \frac{v^2}{c^2}\right)$$

$$= \frac{\ell v^2}{c^3},$$

where the binomial theorem has been used and only terms up to second order have been retained, i.e., v is assumed to be so small as compared to c that the terms involving v^3/c^3, v^4/c^4, etc. can be neglected. This difference in time causes one beam to lag behind the other through a distance d given by

$$d = c \, \Delta t = \frac{\ell v^2}{c^2}. \tag{A-2}$$

Equation (A-2) can be used to determine the velocity v in terms of c, ℓ and t. Unfortunately Δt is so small that it is not possible to measure it experimentally. However, the interference of the two beams of light can be used to quite an advantage. Since the light source S is not confined to a point and has finite extensions, the different rays of light originating from different points of the source are not in phase with one another. In addition, the phase relation between the rays of the two beams of light is affected by the velocity of the earth. Thus on falling on the cross wires of the telescope these rays produce interference fringes. This would be a still fringe pattern and we cannot isolate the effect of the velocity of the earth. If, however, the apparatus is rotated about a vertical axis, the effect of the motion of the earth would produce a change in the phase relation and the interference fringes would move sideways. To get the maximum shift the apparatus was rotated through 90°. If n is the number of fringes shifted corresponding to the path difference 2d, then

$$2\,d = n\,\lambda,$$

which, on using equation (A-2), yields

$$n = \frac{2\,d}{\lambda} = \frac{2\,\ell}{\lambda}\frac{v^2}{c^2}\,. \tag{A-3}$$

The velocity of light, c, is approximately 3×10^8 m s^{-1}. The speed v of the earth is 3×10^4 m s^{-1}. If sodium light is used as a source, λ may be taken as equal to 589 nm. In actual experiment, the effective length ℓ was made equal to 11 metres through the use of multiple reflections. Substituting these values in equation (A-2), we get n = 0.4 fringe. The experimental precision was about 0.01 of a fringe. Therefore a movement of 0.4 of a fringe, on 90° rotation of the apparatus, could be easily detected. To the surprise of Michelson and Morley, no such shift was observed. The experiment was repeated at different places and in different seasons of the year but the result was always the same. Hence contrary to expectations, the speed of the earth could not be detected by performing an optical experiment confined to the earth.

Before Einstein put forward his special theory of relativity, many unsuccessful attempts were made to explain the negative result of the Michelson-Morley experiment. We give a resumé of some of these attempts.

1. The negative result of this experiment was explained by assuming that

the earth dragged the ether along with it so that it was always at rest with respect to the ether and consequently the velocity of light relative to the ether would be the same in all directions. But this was in contradiction with the concept of a stationary ether: an essential assumption for explaining the *aberration of light*. Neither one of these two hypotheses can explain both the experiments.

2. The Michelson-Morley experiment was explained by Lorentz and Fitzgerald who made the ad-hoc assumption that material bodies moving with a velocity v contracted by a factor of $\sqrt{c^2 - v^2}$ in the direction of their motion. This contraction, however, cannot be observed because it applies to all bodies including the measuring rods. For an interferometer in which $\ell_1 \neq \ell_2$, it predicted a change in phase difference with velocity. Such a change was never observed.

3. It was also suggested that Maxwell's equations might be incorrect in their conventional form. The correct equations would be invariant under the Galilean transformation so that the inertial frames of reference could not be distinguished even by optical experiments. In this case it whould not be possible to determine the velocity of the earth by performing an experiment on the earth. However, the "correct" equations could not be formulated: all the modifications of Maxwell's equations yielded results which did not agree with experiment.

APPENDIX B

ELEMENTS OF GROUP THEORY

What is a group? This can best be answered by reference to a few examples.

Consider the set S of elements 1, −1, i, −i, where $i = \sqrt{-1}$; we usually write this set as $S = \{1, -1, i, -i\}$. We note that, with the ordinary multiplication of numbers as the law of composition, the elements of S possess the following properties:

1. If an element of the set S is multiplied by any element of the same set, the resulting number is again an element of S. For example:

$$1 \times i = i, \quad i \times (-i) = 1, \quad (-1) \times (-1) = 1.$$

2. The multiplication is associative, i.e. if a, b, c are the elements of S, then $a \times (b \times c) = (a \times b) \times c$. For example:

$$1 \times \{(-1) \times i\} = 1 \times (-i) = -i$$

and $$\{1 \times (-1)\} \times i = (-1) \times i = -i,$$

so that $1 \times \{(-1) \times i\} = \{1 \times (-1)\} \times i$.

3. The set S contains an element, namely 1, which when multiplied by any element of the same set either from the left or from the right reproduces that element:

$$1 \times 1 = 1, \quad 1 \times (-1) = -1, \quad 1 \times i = i, \quad 1 \times (-i) = -i.$$

Such an element is known as **identity element.** It may be noted that 1 is the only element having this property. It is therefore *the identity element.*

4. To every element of S there corresponds an element of the same set such that the product of the two elements in any order is the identity element. Such elements are called inverses of each other. For instance, as

$$i \times (-i) = (-i) \times i = 1 = \text{identity element},$$

i and −i are the inverses of each other. Similarly 1 and −1 are their own inverses. It may be noted that every element of S has one and only one inverse.

Similarly the set $T = \{1, \omega, \omega^2\}$, where ω is a cube root of unity, can be seen to possess the above four characteristics with respect to the ordinary multiplication of numbers, with 1 as the identity.

Next consider the set M of three matrices.

$$I = \begin{pmatrix} 1 & 0 \\ 0 & 1 \end{pmatrix}, \quad A = \begin{pmatrix} -1 & -1 \\ 1 & -0 \end{pmatrix}, \quad B = \begin{pmatrix} 0 & 1 \\ -1 & -1 \end{pmatrix}.$$

If matrix multiplication is taken as the law of composition, we note that

1. The product of any two matrices belonging to the set M is again a matrix belonging to this set. For instance

$$AA = B, AB = I,$$

$$BA = I, BB = A.$$

2. The matrix multiplication is associative.
3. Since $IA = AI = A$, the unit matrix I serves as the identity element.
4. Since the three matrices are non-singular, the inverse of each one of them exists. Moreover, equations $AB = BA = I$ show that A and B are inverses of each other. Again the unit matrix is its own inverse. Hence the inverse of each member of the set M exists and is itself a member of the set.

This shows that under matrix multiplication the matrices I, A, B also possess these four characteristics.

Finally, we consider the set J of all integers, positive, negative and zero, i.e.

$$J = \{\cdots, -3, -2, -1, 0, 1, 2, 3, \cdots\},$$

and show that it also possesses all the four characteristics mentioned above provided the law of composition is the ordinary addition of numbers.

1. If m and n are any two integers, then their sum m + n is again an integer and therefore must belong to the set J.

2. The addition of numbers is associative, i.e., $\ell+(m+n) = (\ell+m)+n$, where ℓ, m, n are arbitrary integers.

3. Zero, and not one, serves as the identity element as for any integer m, $m+0 = m = 0+m$.

4. Since for any integer $m + (-m) = 0 =$ identity element, the negative of an integer is its inverse. Thus the inverse of every integer exists and is a member of the set J.

The fact that these sets of different elements possess the same four characteristics with respect to some given binary operation (law of composition) suggests that a common name may be given to these sets. Such sets are known as **groups**. Thus a finite or infinite set $G = \{a, b, c, \cdot\cdot\cdot\}$ is said to form a group with respect to a binary operation, usually called multiplication, if

1. the product of any two elements of G is an element of G; then the set G is said to be closed under multiplication and is said to possess a closure property with respect to multiplication
2. the multiplication is associative
3. G contains an element which on multiplication with any element of the set reproduces that element; such an element is called the **identity** of G
4. the inverse of each element of G, defined as an element which when multiplied with the given element yields the identity element, is also an element of G.

Words like "product" and "multiplication" when used in the context of groups do not necessarily have their conventional meaning. The law of combination may be ordinary multiplication, matrix multiplication, ordinary addition or something else; it is customary to use the word "multiplication" for any kind of binary operation. Since in future we shall often have to refer to the elements of the various sets, it is convenient to introduce the symbol '$\in$' which means 'belongs to' or 'is a member (or element) of' or 'is contained in'. Thus '$a \in G$' means that 'the element a belongs to the set G'.

To crystallize our ideas, we recast the definition of a group as follows: A finite or infinite set $G = \{a, b, c, \cdot\cdot\cdot\}$, is said to form a group under a law of composition if

1. for every a,b $\in$ G, ab $\in$ G
2. for every a,b,c $\in$ G, a(bc) = (ab)c
3. there exists an element e $\in$ G, known as the identity element of G, such that ae = ea = a for every a $\in$ G
4. for every a $\in$ G there exists an element $a^{-1} \in$ G, known as the inverse of a, such that $aa^{-1} = a^{-1}a = e$.

Postulate 3 requires the existence of an element e which should serve as the left and the right identity for all the elements of G. Similarly, according to postulate 4, $a^{-1} \in G$ is required to serve both as the left and the right inverse for a $\in$ G. These postulates are in fact stronger than what are needed for the definition of a group and can be replaced by weaker ones which require the existence of only a one sided (left or right)identity and a one sided inverse of each element of G. From these weaker postulates, postulates 3 and 4 can be derived.

Group elements do not necessarily commute. If all the elements of a group commute with one another, the group is said to be **commutative** or **Abelian.**

The number of distinct elements of a group is known as the order of the group. If the order of a group is a finite number, the group is said to be **finite.** Otherwise it is known as an infinite group.

Subgroup

If within a group G there is a collection of elements H which forms a group under the same law of composition as for G, then H is said to be a subgroup of G. It may be emphasized that a subset H of G forms a subgroup of G only if H is a group under the same binary operation as for G.

For example the set H = {1, −1} is a subgroup of G = {1, −1, i, −i} because H and G are groups under the same binary operation (ordinary multiplication), but it is not a subgroup of the additive group N = {• • •, −2, −1, 0, 1, 2, • • •} because H is not a group under ordinary addition.

It may be noticed that G itself and the identity element alone are subgroups of the group G. These are known as **trivial** or **improper subgroups.** All other subgroups are called **proper subgroups.**

We shall now prove an important theorem which determines whether a subset of a group forms a subgroup or not.

Theorem

A subset H of a group G is a subgroup of G if and only if when a, b

$\in$ H, then $ab^{-1} \in$ H.

Proof

(i) The condition is necessary:

Let H be a subgroup of G and a,b $\in$ H. Then b^{-1} and the product of a and b^{-1}, i.e., $ab^{-1} \in$ H, since the product of any two elements of a group is itself an element of the group.

(ii) The condition is sufficient:

Here we have to show that if for any pair a, b belonging to the subset H of the group G, the product ab^{-1} also belongs to H, then H forms a subgroup of G. Since a,a $\in$ H, $aa^{-1} = e$ should also belong to H, i.e., H contains the identity element. Now as e,b $\in$ H, it follows that $eb^{-1} = b^{-1} \in$ H, i.e., the inverse of every element of H is also an element of H. When a, $b^{-1} \in$ H, then $a(b^{-1})^{-1} = ab \in$ H, i.e., H is closed under multiplication. The associativity follows because it holds in G. Hence H is a subgroup of G.

APPENDIX C

CARTESIAN TENSOR ANALYSIS

Tensors have played an important role in the development of differential geometry. They can be employed very usefully to develop relativistic mechanics and electrodynamics. By using tensors, most of the new concepts in mechanics emerge in a natural way and many physical quantities, like momentum and energy, force and power, current density and charge density, find their natural groupings. Many laws of physics can be expressed concisely through tensors. In the following we give a brief account of those concepts of tensor analysis which are relevant to the development of relativistic mechanics and electrodynamics.

Linear Transformation

Consider a Cartesian coordinate system K in a 3-dimensional Euclidean space. Let (x_1, x_2, x_3) be the coordinates of a point P in K. Consider the linear equations

$$x_1' = a_{11} x_1 + a_{12} x_2 + a_{13} x_3 + b_1 \qquad \text{(C-1a)}$$

$$x_2' = a_{21} x_1 + a_{22} x_2 + a_{23} x_3 + b_2 \qquad \text{(C-1b)}$$

$$x_3' = a_{31} x_1 + a_{32} x_2 + a_{33} x_3 + b_3, \qquad \text{(C-1c)}$$

where a_{ij} and b_i, $i, j = 1, 2, 3$ are constants. The variables x_1', x_2', x_3', being linear expressions in x_1, x_2, x_3, may represent the coordinates of P in some other Cartesian coordinate system, say K', in the same space. Then the above equations are said to define a linear transformation from one coordinate system K to another coordinate system K'.

The origin O(0, 0, 0) of the coordinate system K has the coordinates (b_1, b_2, b_3) in K' as can be seen by putting $x_1 = 0 = x_2 = x_3$ in equations (C-1). This shows that the constants b_i give a measure of the separation between the origins of the coordinate systems K and K'. If the origins of K and K' are coincident, i.e., $b_1 = 0 = b_2 = b_3$, equations (C-1) reduce to

$$x_1' = a_{11} x_1 + a_{12} x_2 + a_{13} x_3 \tag{C-2a}$$

$$x_2' = a_{21} x_1 + a_{22} x_2 + a_{23} x_3 \tag{C-2b}$$

$$x_3' = a_{31} x_1 + a_{32} x_2 + a_{33} x_3 . \tag{C-2c}$$

This linear transformation is then said to be **homogeneous**.

Equations (C-1) may also be written as

$$x_i' = \Sigma\, a_{ij} x_j + b_i , \qquad i = 1, 2, 3, \tag{C-3}$$

where sum over j is to be taken for j = 1, 2, 3, and can be expressed in a simpler form by introducing summation convention. According to this convention if a variable index appears exactly twice in a given term, summation is understood to be carried over all possible values of the index. Such an index is called a **summation index** or a **dummy index**. The Σ sign is, therefore, dropped and equations (C-3) take the form

$$x_i' = a_{ij} x_j + b_i , \qquad i = 1, 2, 3. \tag{C-4}$$

A dummy index in a term may be replaced by another, provided the new dummy index does not occur elsewhere in that term. A letter index which occurs only once in a term is called a **free index**. If we also adopt the convention that an equation holds for each value of the free index, then we may drop 'i = 1, 2, 3' and write equation (C-4) as

$$x_i' = a_{ij} x_j + b_i. \tag{C-5}$$

Equations (C-1) can be written in matrix form as

$$x' = A\,x + B, \tag{C-6}$$

where

$$x' = \begin{pmatrix} x_1' \\ x_2' \\ x_3' \end{pmatrix}, \quad x = \begin{pmatrix} x_1 \\ x_2 \\ x_3 \end{pmatrix}, \quad B = \begin{pmatrix} b_1 \\ b_2 \\ b_3 \end{pmatrix}$$

are three column matrices and

$$A = \begin{pmatrix} a_{11} & a_{12} & a_{13} \\ a_{21} & a_{22} & a_{23} \\ a_{31} & a_{32} & a_{33} \end{pmatrix}$$

is a 3 × 3 square matrix. The matrix A is called the **matrix of the linear transformation** or the **transformation matrix.**

If $b_1 = 0 = b_2 = b_3$, then equations (C-5) reduce to

$$x_i' = a_{ij}\, x_j. \tag{C-7}$$

Further, if each of a_{ij} is equal to unity for $i = j$ and zero for $i \neq j$, then we have

$$x_i' = x_i. \tag{C-8}$$

Such a transformation is known as the **identity transformation.**

If the Cartesian frames of reference K and K′ are rectangular, the linear transformation (C-1) is said to be **orthogonal.** Let us find the relations which are satisfied by the a's occurring in an orthogonal transformation. Let P and Q be two points having respectively the coordinates (y_1, y_2, y_3) and (z_1, z_2, z_3) in K and (y_1', y_2', y_3') and (z_1', z_2', z_3') in K′. Then these coordinates are related by the equations

$$y_i' = a_{ij}\; y_j + b_i,$$

$$z_i' = a_{ij}\; z_j + b_i,$$

for $i = 1, 2, 3$. On subtraction, we get

$$y_i' - z_i' = a_{ij}\,(y_j - z_j), \tag{C-9}$$

which is the transformation law for the coordinate differences. Clearly in specifying the positions of P and Q in a coordinate frame, their location is not disturbed and so they remain the same distance apart whatever coordinate system we use to locate them. For rectangular coordinate systems K and K′, this would imply that the expression $\sum_{i=1} (y_i - z_i)^2$ for the square of the distance between the points P and Q as noted in K will retain its form in going to K′,

so that

$$\sum_{i=1} (y_i' - z_i')^2 = \sum_{i=1} (y_i - z_i)^2.$$

Using the summation convention, this equation may be written as

$$(y_i' - z_i')\,(y_i' - z_i') = (y_i - z_i)\,(y_i - z_i).$$

Substituting the expression for $(y_i' - z_i')$ from equation (C–9) in the above equation, we obtain

$$a_{ij}\,(y_j - z_j)\,a_{ik}\,(y_k - z_k) = (y_i - z_i)\,(y_i - z_i)$$

or $$a_{ij}\,a_{ik}\,(y_j - z_j)\,(y_k - z_k) = (y_k - z_k)\,(y_k - z_k). \qquad \text{(C-10)}$$

Here on the left hand side of the equation, we have used different indices j and k because the summations involved are independent of each other, while on the right hand side, for later convenience, the dummy index i has been replaced by k.

If equation (C-10) is to hold for all pairs of points, then we must have

$$a_{ij}\,a_{ik} = \delta_{jk}, \quad j, k = 1, 2, 3, \qquad \text{(C-11)}$$

where δ_{jk} is the Kronecker delta defined by

$$\delta_{jk} = 1, \quad \text{if } j = k$$
$$= 0, \quad \text{if } j \neq k.$$

Hence equation (C-11) is the relation which the a's must satisfy in order that the linear transformation (C-4) may represent an **orthogonal transformation.** Equation (C-11) is said to represent the **orthogonality condition.**

Thus a linear orthogonal transformation connecting the coordinate systems K and K′ is given by

$$x_i' = a_{ij}\,x_j, \quad i = 1, 2, 3 \qquad \text{(C-12a)}$$

$$a_{ij}\,a_{ik} = \delta_{jk}, \quad j, k = 1, 2, 3. \qquad \text{(C-12b)}$$

Problem

Show that the relation

$$a_{ik}\, a_{jk} = \delta_{ij} \qquad \text{(C-13)}$$

also represents the orthogonality condition.

We have seen that, in matrix form, the linear transformation equations in 3-dimensional space can be written as

$$x' = A\, x + B, \qquad \text{(C-6}'\text{)}$$

where

$$A = \begin{pmatrix} a_{11} & a_{12} & a_{13} \\ a_{21} & a_{22} & a_{23} \\ a_{31} & a_{32} & a_{33} \end{pmatrix}.$$

Then the element in the ith row and jth column of the product of A with its transpose A^T is given by

$$(A\, A^T)_{ij} = (A)_{ik}\, (A^T)_{kj} = a_{ik}\, a_{jk} = \delta_{ij} = (I)_{ij}. \qquad \text{(C-14)}$$

Since i and j are arbitrary, we conclude that

$$A^T A = I. \qquad \text{(C-15)}$$

This shows that the transformation matrix of an orthogonal transformation is orthogonal. Thus an easy way to check whether a linear transformation is orthogonal or not is to find whether the corresponding transformation matrix is orthogonal or not.

Moreover, taking the determinant of both sides of equation (C-15) and recalling that the determinant of the product of two matrices A and B is equal to the product of the determinants of the individual matrices A and B, we get

$$(\det A^T)(\det A) = 1.$$

However, as $\det A^T = \det A$, this equation gives

$$\det A = \pm 1. \tag{C-16}$$

That is, the determinant of an orthogonal matrix is equal to either +1 or −1. The orthogonal transformations with the determinants of transformation matrices equal to +1 and −1 are called respectively the **proper and improper orthogonal transformations.**

To see the geometrical interpretation of the coefficients a_{ij} let a point P on the x_1-axis of the coordinate frame K have coordinates (p, 0, 0). Then the coordinates of P in the frame K′ are

$$x_1' = a_{11}\, p$$

$$x_2' = a_{21}\, p$$

$$x_3' = a_{31}\, p.$$

These equations show that a_{ij} is the cosine of the angle between x_j- and x_i'-axes. The orthogonality conditions simply express the orthogonality of the axes in terms of the direction cosines.

We shall now consider some examples of orthogonal transformations.

Consider two orthogonal coordinate systems K and K′ such that x_3- and x_3'- are coincident and x_1'- and x_2'-axes make an angle θ with the x_1- and x_2-axes, respectively as shown in Fig. C-1. The frame K′ can then be thought to have been obtained from the frame K after rotating it about the x_3-axis through an angle θ in the counter-clockwise direction. Let a point P have the coordinates (x_1, x_2, x_3) and (x_1', x_2', x_3') in K and K′, respectively. Then from Fig. C-1, $OL = x_1$, $LP = x_2$, $ON = x_1'$ and $NP = x_2'$. Thus

$$x_1 = OL = OM - LM = ON \cos\theta - RN = ON \cos\theta - PN \sin\theta$$

or $$x_1 = x_1' \cos\theta - x_2' \sin\theta$$

Similarly, $$x_2 = x_1' \sin\theta + x_2' \cos\theta.$$

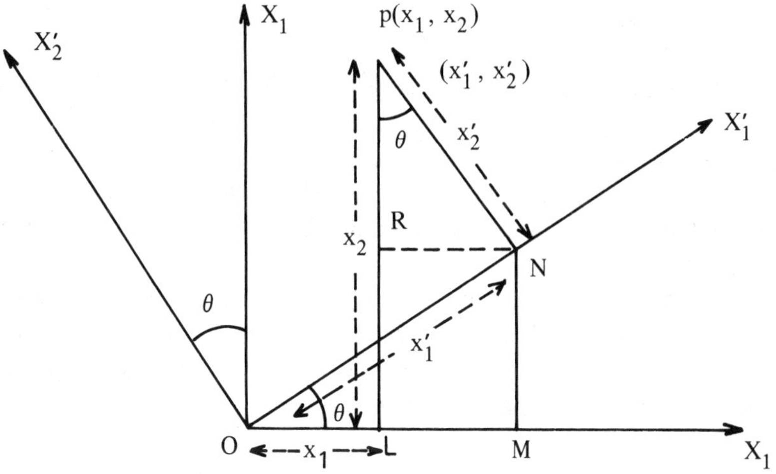

Fig. C-1. Rotation about x_3-axis

The transformation equations for primed coordinates are obtained by solving these equations for x_1' and x_2':

$$x_1' = x_1 \cos\theta + x_2 \sin\theta$$

$$x_2' = -x_1 \sin\theta + x_2 \cos\theta$$

$$x_3' = x_3.$$

The transformation matrix for this linear transformation is

$$A = \begin{pmatrix} \cos\theta & \sin\theta & 0 \\ -\sin\theta & \cos\theta & 0 \\ 0 & 0 & 1 \end{pmatrix}$$

This is a unimodular orthogonal matrix. Hence rotation is a proper orthogonal transformation.

It may be noted that since x_1 and x_2 also represent the components of the vector $\overrightarrow{OP}$, the components x_i of a vector will also transform the same way, viz.,

$$x_i' = a_{ij}\, x_j. \tag{C–7′}$$

We may next consider the same problem with the modification that the origins O and O′ of the two frames of reference K and K′ are not coincident. This will add constant terms, the coordinates of O′ relative to O, to the transformation equations but the transformation matrix A will not be altered. Hence the new transformation is also a proper orthogonal transformation.

Finally, we consider two frames of reference, K and K′, in a three dimensional space such that each is the mirror image of the other. Then the coordinates (x_1, x_2, x_3) and (x_1', x_2', x_3') of the same point in space as noted in these two frames are related by the equations

$$\begin{aligned} x_1' &= -\,x_1 \\ x_2' &= -\,x_2 \\ x_3' &= -\,x_3. \end{aligned} \tag{C-17}$$

This is a linear transformation and is called the **reflection**. The transformation matrix for reflection is given by

$$A = \begin{pmatrix} -1 & 0 & 0 \\ 0 & -1 & 0 \\ 0 & 0 & -1 \end{pmatrix}$$

and is orthogonal. Moreover, since det A $= -1$, the reflection is an improper orthogonal transformation.

Equations (C–17) suggest that on reflection the space coordinates change sign so that $\mathbf{r} \to \mathbf{r}' = -\mathbf{r}$. A vector which changes sign on reflection is called a **polar vector** or merely a **vector**. The momentum **p** also changes sign on reflection and is therefore a polar vector. It may also be noted that on reflection a right handed system of coordinates changes into a left handed system of coordinates. However, the cross product of two vectors **a** and **b** behaves in a different way. On reflection **a** and **b** change their sign: $\mathbf{a} \to -\mathbf{a}$ and $\mathbf{b} \to -\mathbf{b}$. Therefore, on reflection $\mathbf{a} \wedge \mathbf{b}$ remains unchanged. Such a vector is called an **axial vector** or a **pseudovector** or a **vector density**. Angular momentum and magnetic moments are examples of pseudovectors. It may be pointed out that under proper orthogonal transformation **a** and **a** ∧ **b** behave in the same manner. It is only under improper orthogonal transformation that an additional change of sign occurs for a pseudovector.

Physical Quantities: Their Components and Transformation Laws

Physical quantities exist independently of coordinate frames. A coordinate frame provides a convenient way for their description. Such quantities are usually described in terms of their components in a coordinate frame. The number of components of a physical quantity is an integral power of the dimension (usually 3) of the space. These components in different frames are related in a specific way. The number of components and the way they are transformed in going from one coordinate frame to another characterize the physical quantities. In the following we define some physical quantities in an orthogonal coordinate system. The relations between their components are with respect to orthogonal Cartesian coordinate systems having common origin.

We call a physical quantity T a **scalar** if it has $3^0(=1)$ component which transforms under an orthogonal transformation as $T' = T$. We define a **vector** as a physical quantity having $3^1(=3)$ components which transform under the orthogonal transformation (C-12) according to the relation $T_i' = a_{ik} T_k$. Similarly, a **dyadic** is defined as a physical quantity having $3^2(=9)$ components which transform under an orthogonal transformation according to the equation $T_{ij}' = a_{ik} a_{ip} T_{kp}$. Moreover, in mathematics we encounter abstract quantities which in respect of the aggregate of their components and the transformation laws under an orthogonal transformation behave exactly as scalars, vectors, dyadics, etc. and are called Cartesian tensors of rank (or order) zero, one, two, etc. All this suggests the general definition of a Cartesian tensor of rank r as a quantity, *abstract or physical*, having 3^r components which in rectangular Cartesian coordinate systems (or under an orthogonal transformation) transform according to the equation

$$T'_{ij\ldots p} = a_{i\alpha} a_{j\beta} \cdots a_{p\epsilon} T_{\alpha\beta\ldots\epsilon}. \qquad \text{(C–18)}$$

Each component of a tensor of rank r has r letter indices. The number of a's is also r. Like vectors, the indices in tensors refer to the axes. *We shall often refer to the tensor whose components are $T_{ij\ldots p}$ as the tensor $T_{ij\ldots p}$.*

Two tensors $A_{i_1i_2\ldots i_r}$ and $B_{i_1i_2\ldots i_r}$ of the same rank r are called equal tensors if

$$A_{i_1i_2\ldots i_r} = B_{i_1i_2\ldots i_r}$$

for all $i_1, i_2, \cdots, i_r = 1, 2, 3$, i.e., two tensors are equal if the corresponding components of the two tensors are equal.

Two tensors of the same rank are added together by adding the corresponding components of the two tensors. Thus if $A_{i_1i_2\ldots i_r}$ and $B_{i_1i_2\ldots i_r}$ are two tensors of the same rank then their sum $C_{i_1i_2\ldots i_r}$ is defined by relations of the form

$$C_{i_1i_2\ldots i_r} = A_{i_1i_2\ldots i_r} + B_{i_1i_2\ldots i_r} \tag{C-19}$$

in all the orthogonal coordinate systems. If the coordinate systems are related by equation (C-12) then

$$A'_{i_1i_2\ldots i_r} = a_{i_1j_1}\, a_{i_2j_2} \cdots a_{i_rj_r}\, A_{i_1i_2\ldots i_r}$$

$$B'_{i_1i_2\ldots i_r} = a_{i_1j_1}\, a_{i_2j_2} \cdots a_{i_rj_r}\, B_{i_1i_2\ldots i_r}$$

so that

$$C'_{i_1i_2\ldots i_r} = A'_{i_1i_2\ldots i_r} + B'_{i_1i_2\ldots i_r}$$

$$= a_{i_1j_1}\, a_{i_2j_2} \cdots a_{i_rj_r}\, A_{j_1j_2\ldots j_r} + a_{i_1j_1}\, a_{i_2j_2} \ldots a_{i_rj_r}\, B_{j_1j_2\ldots j_r}$$

$$= a_{i_1j_1}\, a_{i_2j_2} \cdots a_{i_rj_r}\, (A_{j_1j_2\ldots j_r} + B_{j_1j_2\ldots j_r})$$

$$C'_{i_1i_2\ldots i_r} = a_{i_1j_1}\, a_{i_2j_2} \cdots a_{i_rj_r}\, C_{j_1j_2\ldots j_r}.$$

The last equation shows that $C_{i_1i_2\ldots i_r}$ is a tensor of rank r.

Let a tensor of rank r have all its components zero. Such a tensor is called the **zero tensor** of rank r. Clearly, its components in all orthogonal coordinate systems are all zero.

Let $A_{i_1i_2\ldots i_r}$ be a tensor of rank r. Let $B_{i_1i_2\ldots i_r}$ be defined by

$$B_{i_1 i_2 \cdots i_r} = -A_{i_1 i_2 \cdots i_r}$$

for all $i_1, i_2, \cdots, i_r = 1, 2, 3$. It can be shown that $B_{i_1 i_2 \cdots i_r}$ is a tensor of rank r. This tensor is called the negative of the tensor $A_{i_1 i_2 \cdots i_r}$. Clearly

$$A_{i_1 i_2 \cdots i_r} + B_{i_1 i_2 \cdots i_r} = 0$$

for all $i_1, i_2, \cdots, i_r = 1, 2, 3$.

It can be seen that if $A_{i_1 i_2 \cdots i_r}$ is a tensor of rank r then, for any real number c, the components $cA_{i_1 i_2 \cdots i_r}$ also constitute a tensor of rank r.

Multiplication of Tensors

Let $A_{i_1 i_2 \cdots i_r}$ and $B_{i_1 i_2 \cdots i_s}$ be tensors of rank r and s respectively. Let $C_{i_1 i_2 \cdots i_p}$, where $p = r + s$, be the set of all quantities obtained by multiplying every component of the tensor of rank r with every component of the tensor of rank s in all orthogonal coordinate systems. Then it can be shown in a straightforward way that $C_{i_1 i_2 \cdots i_p}$ is a tensor of rank $p = r + s$. To prove this for a particular case, for simplicity of notation, let us consider two tensors A_j and B_{mn} of ranks 1 and 2 respectively. Then

$$A'_i = a_{ij} A_j$$

$$B'_{kq} = a_{km} a_{qn} B_{mn} .$$

Let $\quad C_{jmn} = A_j B_{mn} \quad$ and $\quad C'_{ikq} = A'_i B'_{kq}$

Then $\quad C'_{ikq} = A'_i B'_{kq} = a_{ij} a_{km} a_{qn} A_j B_{mn} = a_{ij} a_{km} a_{qn} C_{jmn}.$

This shows that $C_{jmn} = A_j B_{mn}$ is a tensor of rank 3. The proof can easily be extended to tensors of rank r and s.

The tensor of the type C_{jmn} is called the **outer product** of the tensors A_j and B_{mn} and this process of obtaining the new tensor is known as **outer multiplication.** The outer product of more than two tensors can be performed by taking, in succession, products of two tensors at a time.

Contraction of a Tensor

From any tensor of rank r it is possible to obtain a tensor of lower rank $r - 2$ by an algebraic process, known as contraction, described below.

Consider a tensor C_{ijk} of rank 3. Then

$$C'_{ijk} = a_{im}\, a_{jn}\, a_{kp}\, C_{mnp} \,.$$

Set any two free indices equal, say $j = i$ and perform the summation over i. Then

$$C'_{iik} = a_{im}\, a_{in}\, a_{kp}\, C_{mnp}$$

$$= \delta_{mn}\, a_{kp}\, C_{mnp} \,,$$

where we have used equation (C-11). Let us sum over m. Since $\delta_{mn} = 0$ whenever $m \neq n$, all the terms for which $m \neq n$ vanish. The only non-zero term is that for which $m=n$ because then $\delta_{mn} = 1$. The above equation therefore reduces to

$$C'_{iik} = a_{kp}\, C_{nnp} \,.$$

This shows that C_{iik} is a tensor of rank 1. Since only the indices k and p in the tensors C'_{iik} and C_{nnp} are free, we can write D'_k for C'_{iik} and D_p for C_{nnp}. The above equation then takes the form

$$D'_k = a_{kp}\, D_p \,.$$

This process of equating two free indices and performing summation over the repeated index is called **contraction.** A contraction is evidently possible only if the rank of the tensor is either equal to or greater than 2. If the tensor C_{ijk} is the outer product of tensors A_i and B_{ij}, the tensor D_i is called an **inner product** of A_i and B_{ij}. The proof can easily be generalized. Another inner product of A_i and B_{ij} is $A_j B_{ij}$. Thus more than one inner product may be formed from two tensors.

Problem

Show that on contraction the direct product of two vectors yields a scalar.

Solution

Consider two vectors A_i and B_i. The corresponding transformation equations are

$$A'_i = a_{ik} A_k$$

$$B'_j = a_{jp} B_p,$$

which yield

$$A'_i B'_j = a_{ik} a_{jp} A_k B_p.$$

Setting j=i and summing over i, we get

$$A'_i B'_j = a_{ik} a_{ip} A_k B_p = \delta_{kp} A_k B_p = A_k B_k.$$

In vector analysis this is called the **scalar product** of two vectors.

The Kronecker Delta

There does not exist any non-zero vector which has the same components in different coordinate systems. There exist, however, tensors of higher rank which possess the same components in all the frames of reference. Such tensors are said to have invariant components, or frequently, are said to be invariant tensors. Of special interest is the Kronecker delta δ_{ij}. Equation (C-13) can be written as

$$a_{mi} a_{ni} = \delta_{mn}$$

or $$a_{mi} a_{nj} \delta_{ij} = \delta_{mn}$$

If we treat it as a tensor of second rank, then the left hand side of the last equation can be considered as δ'_{mn}. Then

$$\delta'_{mn} = a_{mi} a_{nj} \delta_{ij} = \delta_{mn}.$$

This proves the invariance of the tensor δ_{mn}.

Symmetric and Antisymmetric Tensors

Tensors may have symmetry properties with respect to their indices. As the indices indicate reference to axes of the coordinate system, such symmetry properties are with reference to the description of tensors with respect to directions in space.

A tensor $S_{i_1 i_2 \cdots i_p \cdots i_q \cdots i_r}$ of rank r is said to be symmetric in the pth and qth indices if

$$S_{i_1 i_2 \cdots i_p \cdots i_q \cdots i_r} = S_{i_1 i_2 \cdots i_q \cdots i_p \cdots i_r} \tag{C-20}$$

for $i_1, i_2, \cdots, i_r = 1, 2, 3$. Thus the tensor S_{ijk} of rank 3 is symmetric in the first and second indices if $S_{ijk} = S_{jik}$ for i, j, k = 1, 2, 3. Similarly S_{ijk} is symmetric in the first and third indices if $S_{ijk} = S_{kji}$ for i, j, k = 1, 2, 3. If the tensor is symmetric with respect to all pairs of indices then the tensor is said to be completely symmetric.

A tensor $A_{i_1 i_2 \cdots i_p \cdots i_q \cdots i_r}$ of rank r is said to be antisymmetric or skewsymmetric in the pth and qth indices if

$$A_{i_1 i_2 \cdots i_p \cdots i_q \cdots i_r} = -A_{i_1 i_2 \cdots i_q \cdots i_p \cdots i_r} \tag{C-21}$$

for $i_1, i_2, \cdots, i_r = 1, 2, 3$. Thus the tensor A_{ijk} of rank 3 is antisymmetric in the first and second indices if $A_{ijk} = -A_{jik}$ for i, j, k = 1, 2, 3. Clearly in such tensors, those components for which entries at pth and qth places are equal, are zero. Thus if A_{ijk} is antisymmetric in the first and third indices, i.e., if $A_{ijk} = -A_{kji}$ for i, j, k = 1, 2, $\cdots$, n, then $A_{11k} = 0 = A_{22k} = \cdots = A_{nnk}$ for k = 1, 2, 3. If the tensor is antisymmetric with respect to all pairs of indices then the tensor is said to be completely antisymmetric. Clearly in such tensors all components for which any pair of indices are equal are zero.

Problem

Show that every tensor can be symmetrized or antisymmetrized with respect to any pair of indices.

The symmetry properties of a tensor have an absolute significance, i.e., these are retained in all coordinate systems. To show that, consider, for instance, a symmetric tensor T_{kp} of rank 2 in a frame K. Its components in the frame K′ are given by

$$T'_{ij} = a_{ik}\, a_{jp}\, T_{kp} = a_{ik}\, a_{jp}\, T_{pk} \,,$$

because T_{kp} is a symmetric tensor.

Thus $$T'_{ij} = a_{jp}\, a_{ik}\, T_{pk} = T'_{ji}.$$

We conclude that the symmetry character of a tensor has an absolute significance.

Quotient Theorem

Quotient theorem states that if the product of a given set of quantities A with an arbitrary tensor is also a tensor, then A is itself a tensor. We shall now state and prove this theorem for a particular case.

Theorem

If $A(i, j)\, B_j$ is a tensor of rank 1 where B_j is an arbitrary tensor of rank 1, then $A(i, j)$ is a tensor of rank 2.

Proof

Since the quantities $A(i, j)\, B_j$ and B_j are given to be tensors of rank 1, the transformation equations for them are

$$A'(i, j)\, B'_j = a_{ik}\, A(k, p)\, B_p$$

and $$B_p = a_{jp}\, B'_j,$$

so that

$$A'(i, j)\, B'_j = a_{ik}\, A(k, p)\, a_{jp}\, B'_j$$

or $$\{ A'(i, j) - a_{ik}\, a_{jp}\, A(k, p)\}\, B'_j = 0.$$

But B_j and hence B'_j are arbitrary. Therefore the above relation can hold for all B'_j only if

$$A'(i, j) = a_{ik}\, a_{jp}\, A(k, p),$$

showing that $A(i, j)$ is a tensor of rank 2.

Tensor Calculus

If a tensor is defined at every point of some region in space it is said to represent a **tensor field** in that region. Let us first consider a tensor field of rank zero in a 3-dimensional space, i.e., a scalar field $V(x_1, x_2, x_3)$ in a coordinate system K. We will write it as $V(x_i)$ or more frequently as $V(x)$. If $V'(x')$ represents the same field in K', then

$$V'(x') = V(x). \tag{C-22}$$

Note that here x and x' are the coordinates of the same point in the two frames K and K'. The functions $V(x)$ and $V'(x')$ will not necessarily have the same form but they will have the same numerical value at a fixed point.

Differentiating equation (C-22) with respect to x_k', we get

$$\frac{\partial V'(x')}{\partial x_k'} = \frac{\partial V(x)}{\partial x_i}\frac{\partial x_i}{\partial x_k'}. \tag{C-23}$$

But $$x_i = a_{ki}\, x_k'.$$

Differentiation with respect to x_k' gives

$$\frac{\partial x_i}{\partial x_k'} = a_{ki}.$$

Substituting this value of $\partial x_i/\partial x_k'$ in equation (C-23), we get

$$\frac{\partial V'(x')}{\partial x_k'} = a_{ki}\frac{\partial V(x)}{\partial x_i}.$$

This is the transformation law for a vector. Hence $\partial V(x)/\partial x_i$ represent the components of a vector field. This vector field is called the **gradient** of V and denoted by

$$\text{grad } V \equiv \mathbf{i}_1\frac{\partial V}{\partial x_1} + \mathbf{i}_2\frac{\partial V}{\partial x_2} + \mathbf{i}_3\frac{\partial V}{\partial x_3} = \left(\frac{\partial V}{\partial x_1}, \frac{\partial V}{\partial x_2}, \frac{\partial V}{\partial x_3}\right).$$

The above operation is also called the **gradient operation.**

Let us now consider a vector field characterized by components $F_i(x)$ and $F_i'(x')$ in the coordinate systems K and K'. Then

$$F_k'(x') = a_{ki}\, F_i(x).$$

Differentiating with respect to x_p', we obtain

$$\frac{\partial F'_k(x')}{\partial x'_p} = a_{ki}\,\frac{\partial F_i(x)}{\partial x_m}\,\frac{\partial x_m}{\partial x'_p}.$$

By using equation (C-24), this becomes

$$\frac{\partial F'_k(x')}{\partial x'_p} = a_{ki}\,a_{pm}\,\frac{\partial F_i(x)}{\partial x_m},$$

showing that $\partial F_i(x)/\partial x_m$ is a tensor of rank 2.

In general, if we differentiate a tensor $T_{jk\ldots p}$ of rank r with respect to x_i so that all the letter indices are free, then we get a tensor of rank r+1:

$$\frac{\partial}{\partial x_i}\,T_{jk\ldots p} = R_{jk\ldots p}.$$

This differentiation of the tensor $T_{jk\ldots p}$ with respect to x_i, where i is a free index, is called the gradient of $T_{jk\ldots p}$ with respect to x_i. Such a differentiation enhances the rank of a tensor by unity.

We can contract the tensor $\partial F'_k(x')/\partial x'_p$ by equating the indices p and k and then summing over k. This gives

$$\frac{\partial F'_k(x')}{\partial x'_k} = a_{ki}\,a_{km}\,\frac{\partial F_i(x)}{\partial x_m} = \delta_{im}\,\frac{\partial F_i(x)}{\partial x_m} = \frac{\partial F_i(x)}{\partial x_i}$$

which shows that $\partial F_i(x)/\partial x_i$ is a scalar field, i.e., a tensor field of rank zero. The scalar field $\partial F_i(x)/\partial x_i$ is called the **divergence** of the vector **F** and is denoted by

$$\text{div}\,\mathbf{F} \equiv \frac{\partial F_1}{\partial x_1} + \frac{\partial F_2}{\partial x_2} + \frac{\partial F_3}{\partial x_3} = \frac{\partial F_i}{\partial x_i}$$

The above operation is therefore called the **divergence operation.**

In general, if we differentiate a tensor $T_{jk\ldots p}$ of rank r with respect to x_i, then contract by equating i with one of the letter indices of $T_{jk\ldots p}$ and consequently sum over the dummy index, we get a tensor of rank r − 1:

$$\frac{\partial}{\partial x_i}\,T_{jk\ldots p}\Big|_{j=i} = S_{k\ldots p}.$$

This operation is called the **divergence** of $T_{jk\ldots p}$ with respect to the index j and reduces the rank of a tensor by unity.

We next consider the characteristics of another operation, called the **curl operation,** in a 3-dimensional space. If $F(x) = (F_1(x), F_2(x), F_3(x))$ is a vector field, then by definition

$$\text{curl}\,\mathbf{F} = \begin{vmatrix} \mathbf{i}_1 & \mathbf{i}_2 & \mathbf{i}_3 \\ \dfrac{\partial}{\partial x_1} & \dfrac{\partial}{\partial x_2} & \dfrac{\partial}{\partial x_3} \\ F_1 & F_2 & F_3 \end{vmatrix}$$

The three components of curl **F** are

$$\frac{\partial F_3}{\partial x_2} - \frac{\partial F_2}{\partial x_3}, \quad \frac{\partial F_1}{\partial x_3} - \frac{\partial F_3}{\partial x_1}, \quad \frac{\partial F_2}{\partial x_1} - \frac{\partial F_1}{\partial x_2}.$$

Let us see how these components transform on reflection. Since **F** and **r** are both vectors, on reflection, we have $F_i \to -F_i$ and $x_i \to -x_i$. Therefore

$$\frac{\partial F_3}{\partial x_2} - \frac{\partial F_2}{\partial x_3} \to \frac{\partial(-F_3)}{\partial(-x_2)} - \frac{\partial(-F_2)}{\partial(-x_3)} = \frac{\partial F_3}{\partial x_2} - \frac{\partial F_2}{\partial x_3}.$$

Similarly, other components of curl **F** also remain unchanged on reflection. Hence curl **F** is an axial vector. It is also called **pseudovector** or **tensor density** of rank 1.

Moreover, as F_j are the components of a vector **F**, the quantities $\partial F_j/\partial x_i$ are the components of a tensor of rank 2. Therefore, the quantity

$$\frac{\partial F_j}{\partial x_i} - \frac{\partial F_i}{\partial x_j},$$

being the difference of two such tensors, is itself a tensor of rank 2. We denote it by A_{ij}:

$$A_{ij} = \frac{\partial F_j}{\partial x_i} - \frac{\partial F_i}{\partial x_j},$$

Since the interchange of i and j changes the sign of every component, this tensor is antisymmetric and has only three non-zero independent components. We can choose these three components to be the components of curl **F**. It is for this reason that an operation of the above type is called a curl operation. In general, for a tensor of rank r, the **curl operator** is given by

$$\frac{\partial}{\partial x_p} R_{i\ldots j\ldots k} - \frac{\partial}{\partial x_j} R_{i\ldots p\ldots k}$$

which is a tensor of rank $r + 1$.

Finally, we notice that the operator

$$\nabla^2 \equiv \frac{\partial^2}{\partial x_1^2} + \frac{\partial^2}{\partial x_2^2} + \frac{\partial^2}{\partial x_3^2}$$

is a scalar operator. If it operates upon a scalar field $V(x_1, x_2, x_3)$, we get the scalar $\nabla^2 V$. In general, if ∇^2 operates upon a tensor of rank r, it produces a tensor field of the the same rank.

Tensor Densities or Pseudotensors

We define a tensor density or a pseudotensor T of rank r in a space of n dimensions as a quantity having n^r components which, under an orthogonal transformation, transforms according to the relation

$$T'_{ij\ldots p} = |A| \; a_{i\alpha}\, a_{j\beta} \cdots a_{p\delta} \; T_{\alpha\beta\ldots\delta},$$

where $|A|$ is the determinant of the transformation matrix. For proper orthogonal transformations $|A| = +1$, so that

$$T'_{ij\ldots p} = a_{i\alpha}\, a_{j\beta} \cdots a_{p\delta} \; T_{\alpha\beta\ldots\delta},$$

which shows that under such transformations the tensor densities behave in exactly the same way as the tensors. For an improper orthogonal transformation, $|A| = -1$, so that

$$T'_{ij\ldots p} = -\, a_{i\alpha}\, a_{j\beta} \cdots a_{p\delta} \; T_{\alpha\beta\ldots\delta},$$

which shows that under improper transformations tensor densities do not transform in the same way as tensors. Angular momentum and magnetic moment are the examples of pseudovectors or tensor densities of rank 1.

Tensor Analysis and Laws of Physics

A tensor has been defined as a quantity with a specific number of components which, under an orthogonal transformation, transform according to the relation

$$A'_{ij\ldots} = a_{i\alpha}\, a_{j\beta} \cdots A_{\alpha\beta\ldots}$$

This equation shows that any component of tensor in the primed coordinate system is a linear combination of the components of the same tensor in the unprimed coordinate system, and vice versa, so that if all the components of a tensor vanish in one coordinate system, they must also vanish in every other coordinate system. Now if $A_{ij\ldots}$, $B_{ij\ldots}$ are two tensors of the same rank whose corresponding components are equal in a coordinate system, then

$$A_{ij\ldots} = B_{ij\ldots}$$

$$(A_{ij\ldots} - B_{ij\ldots}) = 0.$$

This shows that all the components of a the tensor (or tensor density) $(A_{ij\ldots} - B_{ij\ldots})$ are zero. Therefore, the components of this tensor (or tensor density) must be zero in every coordinate system. In other words, the equation

$$A_{ij\ldots} = B_{ij\ldots}$$

must hold for all choices of axes. Hence if a physical law can be expressed by a tensor equation, it must retain its form under an orthogonal transformation.

This does not exhaust the whole subject of tensor analysis, but this is all we need to know to understand the theory of special relativity.

REFERENCES

1. A. A. Michelson and E. W. Morley, Amer. J. Sc. **34**, 333 (1887).
2. C. S. Wu *et al.*, Phys. Rev. **105**, 1413 (1957).
3. F. C. Champion, Proc. Roy. Soc. **A136**, 630 (1932).
4. O. M. Bilaniuk *et al.*, Ann. J. Phys. **30**, 718 (1962).
5. O. M. Bilaniuk and E. C. G. Sudarshan, Nature **223**, 386 (1969); Phys. Today **22**, 43 (1969).
6. T. Alväger and M. N. Kreisler, Phys. Rev. **171**, 1357 (1968).
7. V. F. Perepelitsa, Phys. Lett. **B67**, 471 (1977).
8. P. V. Ramana Murthy, Lett. Nuovo Cimento, **1**, 908 (1971).
9. R. W. Clay and P. C. Crouch, Nature **248**, 28 (1974).

INDEX